BIBLIOTHECA LICHENOLOGICA

Herausgegeben von

J. CRAMER

BAND 5

Epiphytic Lichens and Air Pollution in The Netherlands

by

TOKE DE WIT

1976 · J. CRAMER

In der A.R. Gantner Verlag Kommanditgesellschaft

FL - 9490 VADUZ

Printed in Germany
by Strauss & Cramer GmbH, 6945 Hirschberg II
ISBN 3-7682-1059-6

CONTENTS

LIST OF APPENDICES

1. INTRODUCTION AND ACKNOWLEGDEMENTS

In 1958, Barkman published a map of the 'epiphyte deserts' in The Netherlands as part of his doctoral dissertation. This map (see Appendix 1) shows the areas in which few or no lichens of the foliose or fruticose type occur (i.e., the zone lacking lichens and the transitional zone), and is based on his findings during a thorough investigation done in The Netherlands in the period around 1950. In 1968 he published a revised version of this map, based on the same data, on the occasion of the First European Congress on the Influence of Air Pollution on Plants and Animals, which was held at Wageningen (The Netherlands). These publications and many others make the relationship between epiphyte deserts and air pollution abundantly clear (see e.g. Ferry 1973 and Hawksworth 1971).

At the end of 1969, the *Instituut voor Stad en Landschap* in Rotterdam asked Barkman to map the situation at that time, but he was unable to find time for the detailed survey that would be required. An attempt was then made to find a way to have the field work done by volunteers. At a meeting attended by L.J. van Dijk of the *Instituut voor Stad en Landschap*, Barkman, and some members of the *Nederlandse Jeugdbond voor Natuurstudie* (NJN), it was decided that the NJN should perform the survey. This resulted in the formation of a work group, called the *Werkgroep Herkartering Epiphyten Nederland* (WHEN), on the following basis:

Scientific direction:

Prof.Dr. J.J. Barkman, *Biologisch Station Wijster*,
Drs. Toke de Wit, *Rijksinstituut voor Natuurbeheer* (since 1971)

Field work:

Members of the *Nederlandse Jeugdbond voor Natuurstudie*
(see Appendix 2)

Organization:

Rijksinstituut voor Natuurbeheer (Research Institute for Nature Management)

Checking of identifications:

Prof.Dr. J.J. Barkman, Biologisch Station Wijster

Drs. F. Daniels, University of Utrecht

Drs. R. Pos, University of Utrecht

H.F. van Dobben, Nederlandse Jeugdbond voor Natuurstudie

Drs. Toke de Wit, Rijksinstituut voor Natuurbeheer

Financing:

Rijksinstituut voor de Volksgezondheid (National Health Organization)

Rijksinstituut voor Natuurbeheer

In 1970, the survey was launched with an instruction weekend. The initial list of participants included 83 NJN members, only a few of whom performed active work in 1970. During the field work the limitations of the method chosen for the study became apparent, and a number of modifications were introduced in 1971, partially as a result of a meeting of active participants held early in the year.

Starting in the spring of 1971, more time could be given to the guidance of the survey because the present author was appointed to the staff of the Research Institute for Nature Management (RIN) where a large part of her work concerned the WHEN project. In the course of the survey a considerable share of the field work was done by a rather small group of NJN members, a few of whom also surveyed a large part of The Netherlands as part of their doctoral work and/or as student assistant in the RIN.

Appendix 2 gives a list of those who collected data, with indication of the main region(s) they investigated. The field work was completed in the autumn of 1973, and between 1973 and 1976 the analysis of the data was carried out by the author. Klaas Westerbrink and Theo van Laanen assisted in the processing of the data. The results are reported in this thesis.

Acknowledgements

To the many persons who have assisted me in the production of this thesis I would like to express my deep appreciation. Special thanks are due to Prof.dr. J.J. Barkman who, through his continuing interest and stimulating discussions, taught me so much about ecology and especially about epiphytes.

In addition he most kindly made available to me his original field notes containing much unpublished data.

Many thanks to those who participated in the WHEN project. The survey was made possible only through the cooperation of all those who did the field work and whose names are mentioned in App. 3. I am particularly indebted to Maarten Brand, Han van Dobben, Erik van Dijk, Pim van der Knaap and Joost Meltzer for the volume of work they did and the valuable discussions we had.

Mr. K. Slings did field work in Noord Holland with the kind permission of Hoogovens IJmuiden.

Air pollution data were made available by several institutions. A complete list is given in section 7.1. For the fumigation experiments reported on in Chapter 16, the experimental facilities were made available by the Instituut voor Plantenziektekundig Onderzoek, (IPO, Institute for Phytopathological Research) at Wageningen.

To Klaas Westerbrink and Theo van Laanen who assisted with the labor of processing the data, sincere thanks. My appreciation goes also to Dr. Chr. van Leeuwen for his continuing interest in my work.

I wish to thank all those who helped in the preparation of the manuscript. The typing was done by Ms. L.M.C. Reuser, Ms. J.M.M. Bade, Ms. Y.C. Griese and Ms. M. Verwoerd. Much of the final manuscript was typed by Ms. K.M.B. van Loen. The majority of the maps and figures were drawn by H.J.M. Wermenbol and H.M. Korbeeck. Among my other colleagues at the RIN who were of help to me, I wish to mention Ms. Chr. van Beek. Ms. I. Seeger-Wolf translated the manuscript.

Finally, my warm thanks are due to Jan J. Lodder for his stimulating help as well as his critical and original comments.

2. THE OBJECTIVES OF THE WHEN PROJECT

Many studies have been done previously in regions in which few epiphytes occur, a large number of them in cities and some in industrial areas (LeBlanc and Rao 1973a). The descriptions almost invariably concern a 'lichen desert', a region with few or no epiphytes. Such lichen deserts occur in and around the centre of a city or industrial area and are surrounded by zones in which epiphyte richness increases gradually. All of the reports suggest a relationship between the presence of air pollution and the absence of lichens. According to Hawksworth and Rose (1970) the zones (in this case the occurence of certain species) are associated with certain concentrations of sulphur dioxide. A good review of the research done up to 1973 can be found in Hawksworth 1971, 1973. In general it can be said that the studies have usually been limited to a small area. Where very large areas were covered the approach was rather coarse. This is the case, for instance, for the maps of lichen deserts in The Netherlands (Barkman 1958, 1969) and the recently published map of Great Britain (Gilbert 1974). Quantitative data on the occurence of epiphytes in large areas are not yet available.

Rather little is known, too, about differences in the degree of sensitivity of lichens and the specificity with respect to various kinds of air pollution.

All this made it desirable to determine the present distribution of epiphytes in The Netherlands in considerable detail. The objectives of the survey can be described as follows:

- Determination of the areas in The Netherlands where few or no epiphytes occur.

- Comparison of the present epiphyte-poor areas with the situation around 1950.

- Investigation of the relationship between reduced occurence of epiphytic lichens and increased air pollution, particularly to determine whether lichens can serve as an indicator of air pollution and/or which species and inventory methods are to be considered for this purpose.

- Determination of the present distribution of a number of epiphytic lichens and comparison with their distribution in 1950 (and possibly also before 1900).
- Investigation differences in specifity and degree of sensitivity of lichens.
- Providing a basis for further (experimental) research concerning the influence of air pollution on lichens.

3. CONSIDERATIONS CONCERNING THE CHOICE OF THE INVENTORY METHOD

As point of departure, it was decided that data were to be collected for each square (grid unit measuring 5 x 5 km; see 4.2). In the final choice of the inventory method, the considerations discussed below played a role.

3.1 The influence of environmental factors other than air pollution

Since the general objective of the study was to determine what influence a particular environmental factor, i.e., air pollution, has on the distribution of epiphytic lichens, the influence of all other environmental factors had to be excluded as much as possible. This is a difficult problem, because not all of these factors are known and consequently they cannot all be taken into account. Important factors for the distribution of lichens include:

1. The phorophyte

   It is essential to evaluate each square according to tree species, because a much richer epiphytic flora is found in squares with two or more species of trees, for instance both elm and oak, than in those with only one. The reason for this divergence is that to a certain extent different epiphyte species occur on different tree species (see also section 5.1.1). This factor can be excluded by comparing the number of epiphytes found on one tree species (or genus) rather than those on combinations of phorophyte species (i.e., the trees on which the lichens grow). A problem is formed here by the fact that there is no phorophyte that occurs everywhere in The Netherlands. The elm is found mainly in the northern and western parts of the country, the oak on the sandy soils in the central and eastern parts and on the coastal dunes. In the Brabant region it is predominantly the poplar and lime which are suitable for investigation. To obtain a picture of the entire country, the results for the individual phorophytic species must be comparable.

2. The microclimate

The microclimate is extremely important for the epiphytic lichens. In forests, for instance, the epiphytic vegetations differ widely from those occuring on free-standing trees, and different types of forests or woods can also differ strongly in this respect. To permit comparison of all samples (or squares) with each other, the sampling stations should as much as possible have similar microclimates. This can be achieved best by examining only free-standing trees, e.g. trees along roads, driveways, canals and so on. It is also necessary to restrict the sampling to vertical trees, because the slope of a tree can have a strong influence on the microclimate. Strongly shaded trees must also be left out of consideration, and the same holds for very young, slender trees, because the epiphytic vegetation on the latter is usually still in an early stage of development (Gilbert 1968).

3. Climate

Climatic differences within the region in question (in this case The Netherlands as a whole) cannot be excluded by the choice of method for the survey, but this influence must be taken into account in the interpretation of the results. In this respect the drier climate of cities is of special importance. This factor is often put forward, either in addition to or instead of the factor of air pollution, to explain the low epiphyte richness in cities. However, the findings in The Netherlands around 1952 showed clearly that dryness is certainly not the sole factor, because the lichen deserts extended over areas without urban conditions. (See also Coppins 1973 and LeBlanc and Rao 1973b).

The same also holds for other geographical differences such as soil type.

4. Eutrophiation

Eutrophiation with either artificial or natural fertilizers occurs in the vicinity of pastures, farms and so on. This factor has a distinct effect on the epiphyte vegetation. The presence of this factor is difficult to determine; it can usually be detected from the epiphytes, but it is not possible to exclude eutrophiated trees consistently. It is preferable to try to include eutrophiated trees as well as non-eutrophiated trees in all squares; the more so because in the squares with few trees a large percentage are sometimes situated on farms.

There are many other factors with a direct or indirect influence on epiphytes, but in many cases little is known about the influence. A few examples include:

5. Bark wounds: As a result of the presence of wounds in the bark, there is an area on the trunk of the tree with a higher level of nutrients and a higher pH value. This has a distinct direct influence on the epiphyte vegetation.

6. Rain-water tracks, which cause nutrient and moisture gradients.

7. Dust accumulation on the trunk of the tree, particularly at the base and on parts with more slope (e.g., at the origin of branches, slanting trunks). The latter does not play a role if only vertical trunks are used.

8. Structure of the bark (smooth or rough): Grooves give rise to a different microclimate than do smooth surfaces. It has also been found that the surface structure can vary considerably between mature trees of the same species.

3.2 The choice of parameters

For a study performed to find out whether air pollution has an influence on epiphytic lichens, it is necessary to decide which aspects of the lichens are to be investigated. In this case it is taken as working hypothesis that air pollution limits the viability of lichens in some way, and that this influence in not equally strong for all species of lichens. This leads to the following assumptions:

In places where the air is polluted,

a. fewer species of epiphytes will be present (due to the disappearance of sensitive species);

b. the species that occur will be poorly developed;

c. the species that occur will be less fertile; and

d. the percentage cover and the frequency of each of the species present will be lower than in an area with less air pollution.

Thus, the parameters for the condition of the vegetation to be taken into consideration are number of species, vitality, fertility, percentage cover and frequency of occurence.
Here, the species is taken as unit. If a syntaxonomic unit or a life-form is taken as basis, comparable parameters are obtained: number of units, percentage cover and frequency of the units, and composition of the units (the species number; vitality and fertility of the individual species).

The amount of information collected increases with the number of aspects taken into consideration. The investigation of all aspects at the same time is, however, very time-consuming, and the expected amount of information must be weighed against the amount of time it will take to collect it.

3.3 The methods developed by other investigators

Many investigators have reported that epiphyte richness increases outward from a lichen-poor centre. In many of the investigated areas a distinction was made only between a few zones (Hawksworth 1971, 1973). In these cases the method applied was more qualitative than quantitative. The zones concerned are usually:

1. the lichen desert, where only a few (crustose) lichens occur,
2. the transitional zone, sometimes divided into an inner and an outer transitional zone, where the number and vitality of the epiphytic lichens are reduced, and
3. the unaffected area.

Other authors also use species composition for distinguishing zones (Hawksworth and Rose 1970). In these cases zone borders coincide with the appearance of certain species.

In addition to the methods based on the number of species (Gilbert 1965; Pyatt 1970) or the frequency or percentage cover (Domrös 1966), the most important of the quantitative methods developed so far are those based on the index of atmospheric purity (IAP) and comparable indices. Many objections can be raised to the use of these indices, however.
The value used for the sensitivity of a species is the Q given by Barkman in 1958 (de Sloover 1964; de Sloover and LeBlanc 1968) or the mean number of companion species (de Sloover and LeBlanc 1970; LeBlanc and de Sloover 1970) or a coarsely determined value like the P of Trass

(1968, 1973). The Q values like that of Barkman have the disadvantage that although they are expressed by a number, this number indicates only a rank order. For the calculations it is usually assumed implicitly that, for instance, one species with a value of 6 is just as 'good' as two species with values of 2 and 10. When the sensitivity value is derived from the number of companion species, it must be kept in mind that the Q values in this case are also dependent on the relative number of data collected in a poor or a good area. Furthermore, the Q value is dependent on the substrate and on other variables which, in addition to air pollutio have an influence on the number of species present. This severely limits the application of this method. The changes in the equation proposed by Moore (1974) do not solve the problem.

Similar objections hold for the estimation of the percentage cover, wether or not in combination with the frequency, and the numerical signifi cance to be assigned to it. The sensitivity value, moreover, has the disadvantage that it often varies from area to area (Barkman et al. 1969) and also according to substrate (Gilbert 1968).
Therefore, before sensitivity values can be used, the area to be surveyed must be thoroughly investigated. According to LeBlanc (1971), the results based on the IAP are not much more reliable than those obtained with a simpler technique.

4. THE METHOD CHOSEN FOR THE PRESENT SURVEY

4.1 The method initially considered

Initially, it was intended to use the IAP method for the WHEN project. For this purpose, Barkman developed the formula IAP= $\sum_{i=1}^{n} Q \times T$, in which Q is the sensitivity value of a species and T a measure for the cover (transformed value of Braun-Blanquet's scale as modified by Barkman, Doing and Segal). In the field work, 38 species of epiphytes were to be used (see Appendix 3), all of which are relatively easy to recognize in the field. A Q value was assigned to each of these species by Barkman (see Appendix 3 and Barkman 1958).

On the basis of the considerations mentioned in section 3.3, it was decided not to use this IAP or one of the other available quantitative methods. Nevertheless, the data were collected on a basis permitting quantitative analysis. This of course makes it easier to compare the lichen data with data pertaining to air pollution and other factors.

After a short period at the beginning of the survey (see section 5.1.2), the field work was performed only by participants with an extensive knowledge of the species. This made it possible to include more than the 38 species initially chosen. Only a few difficult taxa were excluded in the analysis of the data (see page 32 , Appendix 4).

4.2 The grid

The division of the Netherlands applied (see Appendix 13) was the same as the one used for the floral survey (IVON project) of the Rijksherbarium of the University of Leiden. The large rectangular blocks, numbered from 1 to 62, correspond with the map sheets of the Topografische Dienst of The Netherlands. Each of these sheets is divided into 40 squares of 5 x 5 kilometres. These squares are numbered as follows: the upper row 11-18, that row under that 21-28, and so on (5 rows). The combination of sheet number and square number gives the exact location of a square (e.g., the town of Den Helder lies in square 9.53).
A map with all geographical names used in this publication is given in Appendix 114.

4.3 The method used

The parameters mentioned in section 3.2 were to be considered for use in the WHEN survey; but because of the large number of samples to be taken and the limited manpower it proved impossible to consistently estimate the frequency and coverage or record the fertility and vitality. Thus, the only remaining applicable parameter was the number of species, which many authors consider extremely important.

On this basis, the following conditions were laid down for the survey:

1. All 5 km squares of The Netherlands were to be included.
2. The most important parameter was to be the number of species per unit.
3. The unit to be adopted for the survey was a sample consisting of the epiphytic flora of a group of trees of the same species standing fairly close together.
4. For each sample, a species list was to be made, recording the number and names of the species present.
5. To insure that the microclimate of the trees in the samples was as similar as possible, only the lichens on free-standing, vertical trees and occuring on all side of the trunk up to 2 metres above the ground were included.
6. Whenever possible, supplementary data were to be recorded, e.g. respecting vitality, fertility, abundance and frequency.
7. Wherever possible, at least three samples from one tree species were to be examined in each square.

The following can be added concerning the way in which the survey was carried out:

- When the observer was not sufficiently familiar with the species to make a complete list of all of the lichens present, he confined himself to the 38 species initially chosen (see section 5.1.2).
- In a few cases brushwood was inventoried as well as free-standing trees, but this was usually indicated so that it could be taken into account in the analysis of the data.

- Slightly inclining trees were sometimes included, especially among the willows.
- Very slender trees were avoided.
- Within squares, the distribution of the samples was not always taken into account.
- Stronger eutrophiation of trees near or on farms was not taken as a reason for exclusion, and such trees were included in most of the squares.
- In very many cases the thickness of the tree was indicated.
- The number of trees per sample varies from a few to more than twenty; this number is usually indicated.
- In most cases only the genus of the tree was recorded. The following list indicates the tree species included:

Acer	:	usually _A. pseudoplanatus_, sometimes also _A. platanoides_ or _A. campestre_
Fraxinus	:	_F. excelsior_
Juglans	:	_J. regia_
Populus	:	often _P. canadensis_ of _P. nigra_ but also others
Quercus	:	_Q. robur_ unless _Q. rubra_ is specifically mentioned
Salix	:	usually _S. alba_; other cases usually mentioned
Tilia	:	various species
Ulmus	:	various species

The above species (genera) are always indicated by the generic name. Other species of trees were included incidentally, the most important of these being _Sambucus nigra_ and _Betula_ spp.

- In many cases, for instance almost always for _Lecanora conizaeoides_, mention was made of whether the lichens were found to be fertile or sterile. Usually, only divergent observations were recorded.
- In a few cases the length of fruticose was noted.
- Even when an attempt was made to make a complete list of the lichens present, it remains possible that some species were not observed. In consultation with those who performed the greater part of the field work, a number of taxa were excluded or combined into a single taxon. These taxa are given in Appendix 4.

The foregoing holds for the epiphytic lichens. The mosses, liverworts, and algae were recorded as much as possible but not always consistently. These data are therefore incomplete, and even more so because the familiarity of most of the observers with these groups was rather limited.

5. ANALYSIS OF THE DATA

5.1 Reduction of the data and representation in maps

5.1.1. Mapping of the thoroughly surveyed areas

The aim is to represent the data in such a way that areas with relatively few lichens stand out clearly on the map. The available data consist of species lists per tree species per square; the limitations of the comparability are discussed in Chapters 3 and 4. To avoid over-complication of a map, it is advisable to reduce the amount of information and restrict the number of items included in the legend. To this end, the squares were classified according to the number of lichen species. When choosing the classification it is important to keep in mind that the choice of this classification means a choice of the number of squares in which the epiphytic vegetation is rich or poor.
A problem is, that the classes, which must be established separate for each tree species, must be mutually comparable. The data must be analysed per tree species, because the total number of lichen species in a square with, for instance, both oak and elm, will be greater than in an otherwise completely comparable square where only one of these two tree species is considered or present. This leads to a separate map for each tree species; and to form a comprehensive picture of the whole country, these maps must be combined. This is necessary, because none of the tree species occurs in all squares. The information on lichens will be incomplete unless several species of trees are included.

The following differences between tree species have an influence on the classification:

a. The species number and species composition is different on a different species of tree even when conditions are comparable.
b. The distribution of the tree species over the country is not random. Some trees may occur relatively abundantly in lichen-rich or lichen-poor areas.

The squares were classified on the basis of the number of lichen species per tree species per square. A norm was required for this classification.

Since the separate maps for the tree species were to be combined, the classifications had to be comparable. If under similar circumstances tree species A has x species of lichen and tree species B y lichen species, both tree species must be assigned to the same class when they bear x respectively y lichen species. It would be easiest to make comparable classes if all the tree species could be compared in a homogeneous area under the same conditions, but this is unfortunately impossible. Furthermore, in an area where a tree species occurs only incidental it is often relatively poor in epiphytes, for instance the few elms in the "oak region" (see also chapter 13).

There are several methods by which comparable classes can be obtained. One possibility is to take the average (over all squares) number of lichen species (per square per tree species) as the norm for comparison, but this has the disadventage that the average is strongly influenced by accidental peak values. This objection holds even more for the maximal number of lichen species on a tree species per square.

We therefore preferred to use frequency distributions (total number of squares as a function of the number of lichen species per square). These frequency distributions were calculated after 98 per cent of the data had been collected. The cumulative distributions are given in Appendix 5. The frequency distributions were determined for the tree species sampled in at least 100 squares. The squares were classified into six classes. Four of the six classes comprised 20 per cent of the squares and two classes 10 per cent each. The class-limits lie at the number of lichen species corresponding with 20, 40, 60, 80 and 90 per cent of the number of squares in which the tree species occurs. These limits for the various tree species are given in Appendix 6 (classification T_0).

In this way all squares were classified for each of the tree species examined in the square. For the final combined map, these values had to be combined into a single class per square. For this purpose, it was decided to take as final value the highest class to which a square belonged according to the individual tree species. This was done because what we wanted to find out was how many epiphytes can occur in a square at a given level of air pollution and the lower values could have occurred because other conditions had been less favourable for the tree species with a lower epiphyte-richness. This of course involves the assumption that air pollution was about the same everywhere in a given square.

When using for each tree a classification in which there is the same number of squares in each of the classes, the combined map will show many more squares in the high than in the low and intermediate classes.
To compensate for this to some degree, the highest of the five classes was divided into two classes. This resulted in a better differentiation in the species-richer parts of the country on the combined map, nevertheless there are with this classification also more squares in the species-rich than in the species-poor classes (see Appendix 7 and 11).

From here on, the term epiphyte class will be used to indicate the class assigned to a square on the combined map. Under epiphyte richness is understood the number of epiphytic lichen species of free-standing trees in a square, standardized, surveyed and analysed according to the method applied in this investigation (a high class, a high epiphyte richness means many species). When a tree species is occurring relatively frequently in epiphyte-rich or epiphyte-poor areas this can lead to over- respectively underestimation of the number of lichens observed in a square. As a result, especially in cases of overestimation, a rather species-poor square may be classified relatively high class. (Under-estimation is sometimes compensated for on the combined map, in squares where a different tree species determines the classification).

The following method was used to investigate whether a given tree species occurs relatively more often in epiphyte-poor or epiphyte-rich areas. For every tree species a new combined map was made omitting that species. Thus any incorrect classification for the tree species in question would not affect the new combined map. For each of these maps we counted the frequency of each of the epiphyte classes a) for the whole of the map and b) for those squares in which the trees species occurs. The differences were then tested for significance with a chi^2 test.
If a significant difference is found it is concluded that the classification made on the basis of this frequency distribution is not comparable with those for other tree species distributed differently over The Netherlands.

The comparison of the distribution over epiphyte-poor and epiphyte-rich areas done in this way gave the following chi^2 values:

Acer	11.04	just not significant
Fraxinus	11.15	just significant
Juglans	7.94	not significant
Populus	13.15	significant
Quercus	6.50	not significant
Salix	45.46	significant
Tilia	23.13	significant
Ulmus	21.11	significant

(in all cases there are 5 degrees of freedom, the critical value of chi^2 at the 5% level of significance is 11.1)

Thus, five of the eight species showed a significantly different distribution.

An attempt was therefore made to compensate for these deviations by applying a correction. This was done in the following way. As point of departure, the 20-20-20-20-10-10% classification was taken. If a species occurs relatively more often in epiphyte-poor areas, a relatively larger number of data should fall in an 'epiphyte-poor' class.

Example: For the map resulting from the combination of all tree species except Ulmus, we have:

	a	%	a'	b	$\frac{b}{a'}$	c	$c \times \frac{b}{a'}$	d	d-cumulative
class 1	80	5.93	31.19	50	1.60	20	32.0	29.77	29.77
class 2	206	15.28	80.37	86	1.07	20	21.4	19.91	49.68
class 3	247	18.32	96.36	77	0.80	20	16.0	14.88	64.56
class 4	325	24.11	126.82	105	0.83	20	16.6	15.44	80.00
class 5	195	14.47	76.11	77	1.01	10	10.1	9.40	89.40
class 6	295	21.88	115.09	131	1.14	10	11.4	10.60	100.00
	1,348	100	526	526		100	107.5	100.00	

a =number of squares in the class (total is less than the total number of squares on the map, since squares in which only Ulmus occurs are omitted, leaving 1,348 squares)

b =percentage of total number of squares in which Ulmus occurs

a'=number of squares in which Ulmus occurs, on the basis of the frequency distribution of all squares ($a' = a \times \frac{526}{1,348}$).

c =percentage intervals used for the original classification. Multiplication of c by $\frac{b}{a'}$ increases or decreases this percentage proportionally according to the occurence of the various classes in the squares in which Ulmus occurs. The resulting percentages are brought to the 100%

total, giving the new basic percentages for the classification of Ulmus. With the aid of the cumulative percentages, this new classification can be read directly from the cumulative frequency distribution (Appendix 5) of the number of lichen species on Ulmus.

In the original classification, for Ulmus as for the other species of trees, the 20% poorest squares were called class 1. However, comparison with the other tree species showed that Ulmus was over-represented in class 1, i.e. 50 (b) squares with Ulmus in class 1 instead of the expected 31 (a'). Thus, in the classification relatively more squares must fall in class 1 for Ulmus too: $\frac{b}{a'}$ more than the 20% in the original classification, $c \times \frac{b}{a'}$ is 32% for class 1. On this basis, we recalculated the percentage of squares which, according to Ulmus alone, would fall in each class. The resulting sum does not, however, amount to 100%, and we therefore normalized again to 100% (d). By using d-cumulative, the new class-limits can be read from the cumulative frequency distribution of the number of lichen species on Ulmus. On the original map (T_0) the squares with 0-5 lichen species (which is 20% of all Ulmus squares) belonged to class 1; now, with the new classification, squares with 0-8 lichen species (29.17%) belong to class 1.

In this way the classification was recalculated for each species of tree. This gave the T_1 classification (see Appendix 6), on the basis of which a new combined map was made (Appendix 8).

The entire procedure could now be repeated with the new improved map as point of departure. It was expected that when this procedure had been repeated several times (see Appendix 9), the resulting classification would ultimately show no divergence from the one used at the departure of the procedure. This proved to be the case. Classifications T_1, T_2, T_3 and T_4 were obtained successively (see Appendix 6, 8, 11 and 12 and the scheme in Appendix 10). The T_5 classifications obtained by the application of T_4 (5th round) are all the same as for T_4. The final map was then made on the basis of T_4 (Appendix 12).

The difference between T_4 and T_2 is very small: a square with 7 species of lichens on Quercus belongs to class 3 according to T_2 and to class 2 according to T_4; otherwise there is no difference. Differences from T_1 are slightly larger, although these too are not very great. In comparison with the 'unadjusted' map (T_0; Appendix 7) the new map is slightly less 'favourable', due to a shift toward the lower classes (see also Appendix 11). With respect to the location, shape, and size of the epiphyte-poor and epiphyte-rich areas, the difference is rather unimportant.

5.1.2 The procedure applied in the area in which 38 species were surveyed

Classification T_4 could not be applied for the areas where only 38 species were included in the survey (see Appendix 13), because under the same conditions the number of recorded species was lower, since many species were not taken into consideration. All of the squares for which this was the case are situated in the northern part of the country. For this reason, the northern area, consisting of the squares located on the mainland on map sheets 2, 3, 5-8, 10-13, was chosen as the region for the determination of the relationship between the number of recorded species of lichens with and without the restriction to 38 species. For the squares in this region where all species were included we determined how many species of lichen *would* have been recorded if the survey had been limited to 38 species.

For each species of tree, this limited number that would have been found in case of a 38-species survey was plotted against the number of species actually found (see Appendices 14-20). We then calculated for this graphs the average number of observed species for each limited number of species as well as the average limited number of species for a given observed number of species, and these averages were used to draw the best-fitting line. Calculation of the regression lines was not performed, but in these cases the averages made it possible to draw a best-fitting line with almost the same accuracy. Furthermore, a linear relationship is lacking in many cases. An example worked out for *Populus* is given in Appendix 14, in this case the best-fitting line lies approximately between the two regression lines; data for the other tree species (without regression lines) are shown in Appendices 15-20.

With the aid of these 'correlation graphs' for each tree species the classification can be made on the basis of a limited number of species of lichen. For *Juglans*, too few data were available in this region, and this species was therefore not used in the analysis of the incompletely surveyed squares.

The resulting classifications are shown in Appendix 21; the maps based on the results are shown in Appendices 7, 8 and 12.

5.2 Discussion of the applied method

As already mentioned, there are many factors influencing the occurence of epiphytic lichens. By standardizing the survey as much as possible and by applying corrections during the analysis of the data, an attempt was made to obtain comparability of the epiphytic classes assigned to the squares. Nevertheless, other factors can have caused variation in epiphytic richness. Reference has already been made to factors with geographic variability. These factors include, to mention only a few, climate, soil type, macro- and micro accidentation, density of buildings and trees, forests, hedges, amount of open water, farming practices, roads, and traffic. As far as possible, the probability that these factors play a role was investigated by comparison of their distribution pattern with that of the epiphytic-richness.

Another factor with a potential influence is the natural distribution of the lichen species. If part of the country falls outside the natural distribution area of a species, that species will be absent in that region without air pollution being the limiting factor. This holds in The Netherlands for several species e.g. _Buellia canescens_, which is rather coastal and does not occur in the eastern part of the country (see de Wit 1976), and _Lecanora muralis_, which occurs mainly in the southeastern part of the country (see Appendix 91). For most species, however, the natural geographic limits do not lie in The Netherlands or are much less distinct. Differences in the investigators' accuracy and familiarity with the species could lead to divergence in observed epiphyte-richness. The analysis of the data showed, however, that the epiphyte patterns did not coincide anywhere with the areas of the individual investigators. If any differences are caused by this factor, they are probably very small.

In a small part of the country another problem concerns the local use of the 38 species of the original design for the survey. Altough this limitation was compensated for in the analysis of the data, these results are an approximation of what would actually have been found if a complete survey had been done. Furthermore, one can imagine situations in which the 38 species are not representative of the local assortment. Here too it was found that there was no correlation, between the epiphyte pattern and the borderlines of the '38-species areas'. This can be seen by comparing map sheet 10, which shows the area surveyed for the 38 species, with the surrounding area for which complete species lists were made by a different investigator. The influence of the species limitation is probably small. It is to be expected, however, that within a large '38-species area', differentiation of epiphyte-richness is more difficult than elsewhere. The absence of suitable trees also formed a problem in some parts of the country. In otherwise comparable squares, a smaller number of lichen species will be found in a square where only two samples of one tree species could be obtained than in a square where three or four could be taken. In addition, in squares with many suitable trees the richer places can be chosen, whereas in squares with only a few trees, choice is not possible. In some squares a rather large number of samples could be taken, but from different tree species. In the Veluwe region and a few other places (Biesbosch, the Peel) the number of free-standing, well-developed trees of species included in the survey was so small that it was impossible to obtain a sufficient number of samples, and epiphyte classes were not determined for these squares.

The same problem arose for squares bordering on Germany or Belgium or consisting mainly of water. In some of these squares a number of samples were taken, but mostly from poorly developed, partially shaded or solitary trees or at otherwise less suitable places.

As a result, in some squares epiphyte-richness may be lower than it would have been in more favourable circumstances. In other squares an overestimation may have occured, particularly in areas where the survey was very thorough. The latter may have been the case in the area around Haren (province of Groningen) and for a few squares near Utrecht.

A few areas were surveyed by two investigators. It was possible to compare the results to see whether doubling of the number of samples led to a marked increase in the number of species recorded. The differences proved to be small. The number of species did not become much larger by doubling of the number of samples. Almost invariably, the epiphyte class remained the same or became at most one class higher, the latter usually for cases where the number of species had lain close to the upper class limit.

The results also showed that in most cases two samples per square are usually sufficient in epiphyte-poor areas. In richer areas three samples form the minimum and more are sometimes necessary. In practice, the survey method was usually adjusted to these criteria.

Examination of the final combined map (see Appendix 12), in which an epiphyte class is assigned to each square, shows a few striking features. It is clear that the epiphytic-richness is highest in the northern part of The Netherlands. Epiphyte-poor areas occur in:

a) Rotterdam, The Hague and Amsterdam and the whole area between these cities;
b) Twenthe;
c) Arnhem, Nijmegen and their surroundings;
d) the central part of the province of Limburg;
e) parts of the province of Brabant and of Zeeuws-Vlaanderen.

In the interpretation of the data it must be kept in mind that a single square with a low epiphyte class is not highly indicative. The cause can lie in many different factors. The occurrence of entire areas with more or less the same epiphyte richness is much more informative.

Another point here is that the highest epiphyte class of the separate maps per tree species was always applied for the combined map. When an epiphyte-poor square had even one small protected spot with less air pollution and rather more species, a higher class was indicated for the whole square. This means that the poorer areas on the map are smaller than is actually the case.

The maps of the individual tree species (Appendices 22-29) show comparable patterns: the epiphyte-rich and epiphyte-poor areas coincide, in the main, for the various tree species. The results of the comparison of the tree species in question are discussed in Chapter 13.

6. APPLICATION OF THE DATA FOR A GENERALIZED MAP

In some cases a map is more serviceable if it shows larger areas instead of small units like the many small squares on the combined map (Appendix 12). A more generalized version makes it easier and quicker to locate the 'bad' and 'good' areas. A map of this kind can be made by reducing the number of the classes used for the original map, and the reduction can be done by combining classes, for instance by taking two successive classes together or by putting all 'less good' classes into one class. All squares now belonging to one of the new classes are surrounded by a contour line which encloses areas with the same number and abundance of epiphytes. Because the scale applied here is coarser than the one used for the squares map, isolated squares of a class deviating strongly from the surrounding squares are neglected.

The result of combining classes is shown in the map of Appendix 30, which is based on the total original map (Appendix 12). Classes 1-3 have been combined, and it is clearly evident that almost all of the areas with low epiphyte richness occur in the southern part of the country and all of the epiphyte-rich areas in the northern part. This is made even more clear by Appendix 31, a map which gives very rough indications of the areas in which the trees are rich or poor in epiphytes. On this map local differences are not expressed.

For a generalized map use is sometimes made of the occurrence of various life forms. In that case the areas with only crustose lichens form the lichen desert; the next level is formed by the areas with both crustose and foliose lichens; the area with fruticose lichens as well is slightly better; and the "best" is the area with all these and also lichens of the Usnea type. A recent example of this kind of map is the one of England and Scotland based on data collected by schoolchildren (Gilbert 1974). It is quite simple to make a comparable map of The Netherlands (see Appendix 32). Zone 3 of the British survey (i.e., the zone where grey foliose lichens occur on stone) must be combined with zones 0, 1 and 2 in The Netherlands (i.e., no foliose lichens on trees), because for the latter country no data are available on the occurence of epiliths. It is striking that the areas without foliose or fruticose lichens in England are so much larger than those in The Netherlands. This is probably explained partially by differences between the studies. In The Netherlands almost all of the squares have at least two foliose species, _Physcia tenella_ and

Parmelia sulcata, which are fairly insensitive to air pollutants. Where these species are poorly developed and occur infrequently, they could easily be missed by schoolchildren, whereas it is very unlikely that they were overlooked in the present survey. It is not clear whether in England such species as *Buellia canescens* and *Xanthoria candelaria* were assigned to the foliose lichens.

It is evident that a map based on life forms contains considerably less information than one based on species numbers. Furthermore, not all foliose lichens are more sensitive to air pollutants than crustose lichens or fruticose lichens more sensitive than foliose. To obtain a first approximation of the size and location of epiphyte-poor areas, however, the map based on life forms is extremely useful.

7. COMPARISON OF THE EPIPHYTE RICHNESS WITH THE AVAILABLE DATA ON SULPHUR DIOXIDE IMMISSION

7.1 Origin and mode of presentation of the available data

In The Netherlands a number of institutions measure the immission concentrations of sulphur dioxide, other substances (fluorine, nitrous oxide) are measured as well in only a few places. For the present investigation, an attempt was made to collect as many as possible of the data gathered in the years of the survey and a few years prior to that period, i.e., 1970-1972. Since such data are not handled collectively anywhere, it was necessary to approach all of the relevant institutions individually. All of them supplied the requested data if available (in some cases no measurements had been made in the years in question or the data for those years had not yet been processed).

Data were obtained from the following institutions:

Gemeentelijke Keuringsdienst van Waren of Rotterdam (data for Rotterdam and its immediate environs)

Openbaar Lichaam Rijnmond, Centrale Meld- en Regelkamer in Schiedam (data for the Rhine estuary region)

Bureau van Statistiek of the municipality of Amsterdam (data for Amsterdam)

Centrale Coördinatiecommissie Luchtverontreiniging Noordzeekanaalgebied in Haarlem (data for the Noordzee canal region)

Coördinatiecommissie voor de Luchtverontreiniging in Zuid-Holland in 's-Gravenhage (data for Rotterdam, the Rhine estuary, Westland area, 's-Gravenhage, the islands in the province of Zuid-Holland, Delft and Dordrecht)

Gemeentelijke Bouw- en Woningtoezicht of 's-Gravenhage (data for 's-Gravenhage)

Instituut voor gezondheidstechniek - TNO in Delft (data for Delft, its environs, and the Sloe area)

N.V. Kema in Arnhem (data for various places in the country recorded near power stations)

Provinciale Waterstaat Limburg in Maastricht (data for the province of

Limburg)

Gemeentelijke Dienst voor Bouw- en Woningtoezicht of Eindhoven (data for Eindhoven)

Rijksinstituut voor de Volksgezondheid in Bilthoven (data for Twente, Zeeuws-Vlaanderen, Noord-Brabant and Amsterdam)

The results of the measurements are given as 24-hour average concentrations and sometimes also as hourly average concentrations. In most cases the method used for the measurements was not indicated. Although different techniques can give differences in the measured concentrations, it is assumed in the following that such differences are negligible.

In studies done in Great Britain on the relationship between epiphytes and SO_2 concentrations (Hawksworth and Rose 1970; Gilbert 1968) use was made of average SO_2 concentrations for the whole year or only the winter period. In the present study average concentrations at most of the sites could not be obtained or calculated. In many cases only the frequency distribution of the daily values was available. Therefore, we used the median instead of the average (see also Chapter 17.1).

Almost all of the above mentioned institutions convert their data to cumulative frequency distributions. Sometimes these are plotted on logarithmic probability paper, which usually gives a linear curve. This method is used because the norms advised by the Gezondheidsraad (national health council) and TNO are based on frequency distributions.

.2 The analysis of the sulphur dioxide data

For all of the available measuring sites, the cumulative frequency distribution was plotted on log-probability paper. The data concerned were the 24-hours averages (to be called the day values) for the summer (April-September) and winter (October-March) periods separately and for whole years comprising one summer plus one winter period (April-March). In some cases all of the data were available in this form, in others they had to be converted from the day values.

The data for the winters of 1970/1971 and 1971/1972 were available for most of the measuring sites. For the province of Noord-Holland the data for 1971/1972 were lacking, but those for 1968/1969 an 1969/1970 were available.

Three percentile values were read from the graphs prepared as described above:

50 percentile: the concentration below which 50% of the day values lie
75 percentile: the concentration below which 75% of the day values lie
98 percentile: the concentration below which 98% of the day values lie

One of the reasons for choosing the 50 and 98 percentiles was that these values were applied for the Gezondheidsraad/TNO norms. The 50 percentile also corresponds approximately with the average concentration used elsewhere. The 98 percentile is a measure of the 'peak pollution'. The 75 percentile was taken as the value lying intermediate between the 'averag SO_2 concentrations' and the 'peak pollution' at a given measuring site. An example of the type of graph in question can be found in Appendix 32.

In this part of the study use was made only of the winter concentrations. This was done because precise comparison of the pattern of the concentrations of the various percentiles in different seasons, or throughout the year, with the pattern of the epiphytes in Zuid-Holland, showed that the agreement was best for the values of the winter period (see Chapter 10). Other investigators (Hawksworth and Rose 1970) are also of the opinion that the winter SO_2 concentrations show the best correlation with the epiphyte richness.

For comparison of the epiphyte classes of squares with the SO_2 leve we had to know the latter per square. In many cases several measuring sites were located in one square and data were available for several years. The analysis of these data, was based on the following reasoning: The final epiphyte class assigned is based on the tree species with the highest epiphyte richness in that square. It is therefore more probable that the final epiphyte class will correspond with the 'cleanest' measuring site in a square than with a 'less clean' site. If epiphytes are limited by high SO_2 concentrations, the epiphyte richness will show the best correlation with the season having the highest concentrations. Therefore, for each square the measuring site with the lowest concentrations was taken, and use was made of the concentrations for he 'dirtiest' winter period measured at this site. In a few cases, on this basi the various percentile values conflicted with each other. Under these circumstances, we assigned more value to the (almost always agreeing) 50 and 75 percentiles than to the 98 percentile. On this basis we obtained a 50, 75, and 98 percentile value for 100 squares.

'.3 The relationship between sulphur dioxide concentrations and epiphyte richness

'.3.1 Introduction

Sulphur dioxide is generally considered the most important cause of :he occurence of epiphyte-poor areas (Hawksworth 1973). Several authors ıave demonstrated a relationship between the scale they applied for the ·piphytes and the SO_2 concentrations in the air. Most of these cases con- .erned average concentrations, sometimes for the whole year (Gilbert 1968), ometimes for the winter period (Hawksworth and Rose 1970). Such epiphyte cales have usually been designed to reach conclusions about local SO_2 oncentrations on the basis of evaluation of the local epiphyte vegetation for a discussion of such applications, see Chapter 17).

In the present study the question raised was whether a relationship xists in The Netherlands between the epiphyte richness and the SO_2 level nd, if so, how close this relationship is.

Data on SO_2 levels were available for only 100 of the 1,395 squares nto which the country was divided (see sections 7.1 and 7.2; for the ocation of these squares see Appendix 44). The 50, 75, and 98 percentiles alculated by the above described method were compared with the epiphyte ichness. The arithmetical mean was not used, because this would in most ases have involved a great deal of extra work and/or the basic data were ot always available. Instead, the 50 percentile (the median value) was sed. With application of the median, extreme values have less influence than with the arithmetical mean.

.3.2 Comparison of the percentile values with the epiphyte richness

For each square, the percentile values were plotted against the piphyte class of the relevant square (see Appendix 34-36). The 50 per- entile shows a relationship (albeit rather limited) with the epiphyte lass of a given square, particularly at high SO_2 concentrations. At oncentrations higher than 80 $\mu g/m^3$, the highest epiphyte class is class 2. or the 75 percentile there is also a rather distinct correlation at ower SO_2 concentrations and higher epiphyte classes. The same holds for ıe 98 percentile, but here there is considerable spread.

The question is now whether the epiphyte richness is limited primarily by the peak SO_2 values or by the median concentrations. In other words, with which percentile of the SO_2 concentrations does the epiphyte richness show the best correlation? Here, it is possible to calculate correlation coefficients in the usual way, because the epiphyte classes represent not numerical values but only a rank order. We therefore took the following approach to this question.

On the basis of the SO_2 concentrations of a given percentile, it can be 'predicted' to which class the epiphyte richness of a square belongs. For example, if the 75 percentile is 130 $\mu g/m^3$ or higher, the epiphyte vegetation will belong to class 1. The limiting concentrations are chosen visually such that the prediction will be as reliable as possible. Appendix 37 gives the limiting values for the 50, 75, and 98 percentiles.

Nest, a diagram is made (for an example, see Appendix 38) showing the real situation. For the 75 percentile, for instance, there are 18 squares in which the concentration is more than 130 $\mu g/m^3$ and which can therefore be expected to have epiphyte class 1. For 13 of these squares this is indeed the observed epiphyte class, the other five are of class 2. In this way it can be determined in how many of the squares the prediction was correct and in how many squares the predicted class differs by 1,2, 3 or 4 classes from the true (=observed) class.

The result of the predictions is shown in Appendix 39. The degree of accuracy of the predictions is partially dependent on the size of the classes chosen on the basis of the SO_2 concentrations: if very many squares are predicted to fall in class 1, it is more probable that many of the squares actually belonging to class 1 were correctly predicted. Therefore a comparison was made between the observed and randomly expected scores, the resulting chi^2 being a measure of the improbability of the former and therefore of the correlation between SO_2 levels and epiphyte richness (for example, see Appendix 38). Comparison of the 50, 75 and 98 percentiles (see Appendix 39) shows that the 50 and 75 percentiles give the highest number of correct predictions; the chi^2 of these percentiles is also the highest. The 98 percentile gives, however, slightly fewer very large errors (deviation by 3 or 4 classes), but has relatively few faultless predictions. Since there is strong correlation between all of the percentiles, it is hardly surprising that the correlation with the epiphyte levels is not strongly divergent. Nevertheless, the question remained as to whether the correlation could be improved by combining two percentiles.

Combination of the percentiles was done in various ways (see Appendices 40-42). First, lines were drawn perpendicular to the axes at the limiting concentrations (see Appendix 37). This will be called combination I. The limiting concentrations are the same as those used for the individual percentiles except for the combination of the 50 and 98 percentiles, where a slightly different limiting value for the 50 percentile (in combination with the 98 percentile) permitted a better prediction. All of the combinations gave approximately the same results, but they were not better than the individual percentiles.

Another obvious way to combine the percentiles (combination II) is shown in Appendix 43. Because there is a strong correlation between the percentiles and extreme combinations do not occur, this method does not differ greatly from the first. It leads to an increased number of predictions deviating one class and a decreased number of large errors. In particular, 50/98 II shows to have a good correlation with the epiphyte richness.

Combination of all three percentiles was not applied. This would have required complicated procedures, and some orientational calculations suggested that it would not be productive.

The results led to the conclusion that especially the median SO_2 concentrations and the 75 percentile show good correlation with the epiphyte richness. When the 50 or the 75 percentile is combined with the 98 percentile as combination II, the correlation is still better.

8. COMBINED INFLUENCE OF SULPHUR DIOXIDE AND OTHER FACTORS

8.1 Influence of proximity to the coast

Appendix 4 shows, on the basis of the 50/98 II results (see 7.3.2), the divergence between the predicted and observed epiphyte classes. It is striking that particularly in the squares situated along the coast the epiphyte classes are higher than would be expected from the local SO_2 concentrations. How is this to be explained? Evidently, the effect of SO_2 is compensated for by other factors. The coast is not less polluted than inland regions, because the prediction was based on recorded SO_2 levels. It would seem that the same concentrations are less effective along the coast than elsewhere.

This might be explained as follows. Part of the time wind comes from the west, i.e. from the sea. For coastal squares, west winds are 'clean' winds: sources of sulphur dioxide lie to the east but not to the west. Unfortunately, 'compasses' for SO_2 concentrations like the one shown in Appendix45 are only available for a small number of measuring sites and then only for the average concentrations.

In the climate prevailing in The Netherlands, and certainly in the winter, winds from western directions are accompanied by relatively large amounts of rainfall and higher temperatures. With easterly winds there is little precipitation and the winter temperatures are lower (Braak 1950). Lichens are physiologically more active under moist than under dry conditions (e.g. Farrar 1973) and it has also been shown that under wet conditions they are more sensitive to SO_2 (e.g. Türk et al 1974). This means that the epiphytic lichens occuring in the western part of The Netherlands are in the winter physiologically the most active and have the highest sensitivity to sulphur dioxide in the presence of westerly winds. Thus, for the lichens on the coast the physiologically active period coincides with relatively clean air. This does not hold for inland regions. It may therefore be expected that at the same frequency distribution of SO_2 concentrations the influence of SO_2 will be lower for coastal than for inland lichens.

The degree to which the effect of SO_2 is lower is dependent on the degree to which western winds are 'cleaner' than those from other directions, and this in turn is dependent on the extent to which sources of SO_2 occur to the west of the squares in question. The main coastal effect lies in the lower activity of the peak concentrations in the coastal squares, because these peak concentrations coincide with dry wether.

We attempted to incorporate the coastal effect into the calculations as follows. For each square and assuming that all built-up areas emit sulphur dioxide, we determined, on the basis of topographic maps on a scale of 1 : 50 000, the percentage of the area west of the square, (within the 120° located between 60° S of W and 60° N of W, i.e., between SSW and NNW) that is not built up. The percentage of open land was taken as a measure of coast-effect reduction of the SO_2 concentration. Calculation of this reduction was done as follows.

If y is the number of degrees without buildings between 60° N and 60° S of W and x is the reduction factor, it holds that:

$$x = 0.6 \ \frac{y.100}{120} = 0.5y$$

The value 0.6 was chosen arbitrarily to obtain a reasonable reduction.

The reduction was applied as follows. In each square for which x was larger than zero, the 98 percentile was reduced by x%, the 75 percentile by 0.5 x%, and the 50 percentile was not reduced. For concentrations so low that according to 50/98 II and 75/98 II epiphyte class 4 or higher would be assigned to the square, only half of the reduction was applied, since when SO_2 levels are low the coastal effect is probably small.

In addition to the above coast-effect rule, a simplified rule was applied for purposes of comparison i.e., for all squares lying within 10 km of the coast the predicted epiphyte class is one class better than was predicted on the basis of the SO_2 levels (except for squares falling in class 4 or higher on the basis of 50/98 II and 75/98 II).
With the new SO_2 concentrations, corrected for the coastal effect, the method described in section 7.3.2 was applied to determine the correlation with the epiphyte richness.

The new calculations made for the 75 and 98 percentiles and combinations II showed a much stronger correlation. Especially the 50/98 and 75/98 combinations were greatly improved. With 50/98 II, the 'prediction' of the epiphyte class was now not incorrect by more than one class in 90% of the squares (see Appendices 39, 46, and 47). The simplified approach to the coastal effect gives much less improvement than the 'calculated coast effect' (see Appendix 39).
After incorporation of the coastal effect, divergence occurs mainly where the epiphyte level is lower than would be expected on the basis of the sulphur dioxide level (see Appendix 47).

8.2 Influence of population and industrial densities

In the same way as described above, the percentage of built-up land and the number of inhabitants in each square were plotted against the epiphyte class. The population data were taken from a map prepared by the Central Bureau for Statistics (C.B.S.) showing the distribution of the Netherlands population on the basis of the 1960 census. The size of built-up areas was estimated on the basis of topographic maps on a scale of 1 : 50,000 (prepared in the period 1966-1971).

Correlation with population density proved to be very low. The only conclusion that can be drawn is that when there are more than 60,000 people in a square, the epiphyte class of that square is 3 or lower and is at most class 2 when there are more than 140,000 inhabitants.

The results for the built-up areas showed a similar picture (see Appendix 49): above a certain percentage the epiphyte class remains below a certain value, and below that percentage all classes are possible. Here, the borderlines lie as follows:

$>$ 50% built up = highest epiphyte class is 4
$>$ 55% built up = highest epiphyte class is 3
$>$ 65% built up = highest epiphyte class is 2
$>$ 70% built up = highest epiphyte class is 1

These results do not show, however, that degree of urbanization is indeed the limiting factor. For the squares where low epiphyte richness is limited according to the above by population or urban density, the same is valid according to the sulphur dioxide levels (see Appendix 49). Because the correlation between SO_2 and epiphyte level is much stronger, SO_2 is probably the determining factor. Thus, the effect of urbanization, and high population densities is an indirect one, i.e., via the high SO_2 level occurring under these conditions.

But it is also quite conceivable that other factors have a limiting influence on the epiphyte richness in densely built-up and populated areas for instance an unfavourable microclimate, dust and salt near roads, pollutants released by vehicles.

Two squares form an exception to the above-mentioned rules. One of these is the square in which the city of Nijmegen occurs and which shows a high degree of urbanization, a high population density, a low SO_2 level, and a low epiphyte richness. It is concluded that in this square the SO_2

level is probably not typical of the survey stations, for instance because the SO_2 and lichen sampling sites do not coincide or because a higher SO_2 level prevailed during an earlier period. The second square covers the city of Groningen, with a high density of buildings and population, a low SO_2 level, and a high epiphyte class. It is concluded that there must be places in this square where the SO_2 levels are low and the epiphyte level high: i.e., here, urbanization does not have a limiting effect.

The combination of SO_2 levels with population and/or building density does not improve the correlation with the epiphyte richness. This is in accordance with the opinion of most lichenologists, who consider SO_2 as the main factor limiting lichen flora and vegetation in urban areas, in contrast to the "Drought Hypothesis" of Rydzak (see Coppins 1973).

8.3 Discussion

With the approach described in Chapters 7 and 8 it proved possible, to predict the epiphyte richness retrospectively on the basis of a number of variables. In 56% of the 100 squares investigated, the predicted epiphyte class is the same as the observed class, and in 90% of the squares the difference is not more than one class. The agreement between prediction and observation is very high, especially in view of the rather rough scale (5 x 5 km squares) applied. As closer scrutiny also shows (see Chapter 10), there is considerable variation within squares with respect to both the epiphytes and the SO_2 levels. In many cases the SO_2 measurements and the lichen sampling were not done at exactly the same sites, and this could lead to differences between the predicted and observed epiphyte classes. The use of intervals also means that, for instance, a square could be just outside epiphyte class 3 while the predicted class is just within it, and therefore in such cases the resulting difference (of 1 class) has little significance.

The chi^2 value of the predicted and observed epiphyte class is very high (see Appendices 38 and 39). The probability that such agreement is a matter of chance when there is no relationship between the predicted and the observed class is very much smaller than 0.001. Therefore, there is no reason to reject the assumptions made for the calculations.

It can be seen from Appendix 48 that the epiphyte classes were better than predicted only in a very few places. In a number of places, however, the epiphyte richness was lower than would be expected with the variables used so far. Apparently, factors with a negative influence on

×

epiphytic lichens had been omitted, and these factors seemed to concern industrial areas (Rotterdam - Dordrecht, Amsterdam, Twenthe and the Sloe region). Unfortunately, the lack of quantitative or suitable qualitative data on industry made it impossible to introduce new variables for prediction. Industrial activity varies widely in character and intensity, but an influence of fluorides, hydrocarbons, and/or synergistic effects of various substances seems possible. Another possibility is that in these squares the SO_2 level was at one time much higher and the lichens have not yet recovered (see also Chapter 17).

9. EPIPHYTE RICHNESS IN RELATION TO VARIOUS FACTORS

9.1 Industry and population density

It is known from the literature (see e.g. Ferry et al 1973) that urbanization and industrialization are accompanied by low epiphyte richness. Although the survey determined the epiphyte richness for all of the squares, it is difficult to find similar data for industrialization. Various aspects could be used as a measure of 'industrialization', e.g., the number of industries, the number of employees, energy consumption by industry, and so on, but such data were not available per square.

Only the population data of the 1960 census could be converted into number per square (see also section 8.2). The same census also provided the number of persons employed in industry in the various municipalities (Atlas van Nederland, sheet no. XV-1, 1968), and these numbers serve as a measure of the industrial activity in a region, especially since computers were counted as employees in the municipality in which they worked. The number of industrial employees will only be correlated to a certain degree with the polluting effect of industry, because industries requiring much manpower can be relatively 'clean' as far as air pollutants are concerned and low-manpower industries sometimes emit large amounts of sulphur dioxide or other pollutants. We also determined from the topographical map which squares had two or more factories. The map in Appendix 54 shows the squares with factories and/or many inhabitants.

Appendices 50-53 show in various ways the available data on industry and population in relation to the epiphyte richness. For this purpose we used the Lorenz graph (Pen 1970), in which cumulative percentages are plotted against each other. According to Appendix 50, for instance, 80% of the squares with 25,000 or more inhabitants belong to classes 1-3, even though these classes represent only 34.5% of the total number of squares. The more a curve diverges from a straight line, the less proportionally the factor is distributed over the squares. For comparison, Appendix 53 gives comparable graphs for the factor SO_2. The percentages per epiphyte class here differ from those in Appendix 11, because this factor was only investigated for 100 squares. For mutual comparison of the factors, the percentage of squares of a given category falling within the epiphyte-poorest half of the squares is shown.

It was found that when the number of inhabitants and/or the number of employees is very large, the epiphyte level is almost always low.

The negative effect on the epiphyte richness of 25,000 residents in a square and/or 5,000 industrial employees in a municipality is slightly stronger than that of SO_2 levels which according to 50/98 II lead to classes 1-3. It is also clear that the effect of 25,000 inhabitants in combination with more than 35,000 industrial employees is stronger than either alone. It is only probable that the squares with > 25,000 inhabitants and 35,000 industrial employees have more residents than the squares with > 25,000 and < 5,000 employees. On the basis of the a ilable data, these factors cannot be separated. In the 100 squares for which the SO_2 levels are known, most of the squares with > 25,000 residents again fall in categories 1-3. Where this was not the case, the SO_2 levels were in agreement with the observed epiphyte class. This means, as already mentioned, that the effect of population density is probably indirect.

9.2 Soil type

Comparison of the epiphyte map and a soil map of The Netherlands shows no correlation between soil type and epiphyte richness. On most types of soil both high and low epiphyte richness occur, soils of cut-over peats, however, usually show an intermediate richness (mainly in the eatern part of Drenthe). This is, apart from any effect of air pollution, probably due to the type of landscape. The landscape lacks variety here, and these relatively recently reclaimed areas make the same impression as the picture resulting from the redistribution of farm acreage.
Under these conditions there are few well-developed lanes of trees, groups of trees, and the like, which makes the chance of encountering a rich epiphytic flora at standard sampling sites very small.

In Zuid-Limburg the loess region is somewhat more accidented than the area to the north, and it is quite possible that the hilly structure of this landscape gives rise to large differences in pollutant levels and in the (micro)climate. As a result, the chance of high epiphyte richness increases. This region includes the only (small) part of the phyto-geographic and climatological submontane region lying within The Netherlands (Barkman 1958).

A direct relationship between soil type and epiphyte richness is not to be expected. The type of soil can have an indirect effect, for instance via blown-up dust, landscape structure, or characteristics of the phorophyte that can change under the influence of soil type (Barkman 1958). See also chapter 13.

9.3 Climate

The epiphyte richness was also compared with various aspects of the climate prevailing in The Netherlands. For this purpose, use was made of the *Klimaatatlas van Nederland* (1972). It was found, however, that none of the aspects treated in this work could be related to the picture of the epiphyte richness in The Netherlands. The distribution of lichen species sometimes shows a clear relationship with the climate (see also Chapter 14), as does the distribution of syntaxonomic units (Barkman 1958). But the wide differences in epiphyte richness in The Netherlands cannot be shown to have a relationship with climate. If such a relationship exists, it is evidently overshadowed by the influence of other factors.

9.4 Other factors

There are many other factors that can be assumed to have an influence on epiphyte levels (see also section 5.2), but none of them can explain the differences on a country-wide scale, even though they undoubted exert an influence locally. There were, however, no adequate data available to permit investigation of this point. The method chosen for the survey was not designed for such purpose; to the contrary, it was designed to avoid the effects of these factors as much as possible.

10. DETAILED ANALYSIS OF DATA IN THE WESTERN PART OF THE NETHERLANDS

10.1 Introduction

During the WHEN study, much attention was paid to the western part of the country, particularly the island of Voorne and the surroundings of Rotterdam and 's-Gravenhage. At the beginning of the project data were collected in this region by H.F. Dobben, W.O. van der Knaap, and several others, and later an intensive survey was performed by A.M. Brand. The province of Noord-Holland was surveyed by J.A. Meltzer. Supplementary data were collected in the area between these two regions by the author. The work done by Brand and Meltzer was done as part of their graduate studie in biology at the universities of Leiden and Amsterdam, respectively, and has been published in separate reports (Brand 1976; Meltzer 1976).

In view of the great amount of data on both epiphytic lichens and SO_2 levels, it was considered worthwhile to perform the analysis of the data for the western part of the country not only for the individual squares but also for each sampling or measuring site. A survey map of this region is shown in Appendix 55.

10.2 The data

For almost all of the lichen species observed, a map was made showing the sampling sites in this region at which the given species occurred (see also Appendices 65-69). The SO_2 levels were known for 152 measuring sites in this region (see also sections 7.1 and 7.2). In Zuid-Holland the data concern the summer period of 1970, 1971, and 1972 and the winter period of 1970/1971 and 1971/1972, in Noord-Holland the period from the summer of 1968 through the summer of 1970. Data on all of the half-year periods were not available for all sites, since those for the islands in Zuid-Holland before the winter of 1971/1972 were lacking. The location of the SO_2 sites is shown in Appendix 56.

For the analysis of the data, use was made of the SO_2 concentrations corrected for the coastal effect (see section 8.1). The SO_2 concentrations are indicated on maps, for each year or half year a map was made for each of the three percentiles. Similar maps were made with the highest winter, summer, or annual concentration measured, again for all three percentiles.

10.3 Combination of SO_2 and lichen data for Zuid-Holland

Because the number of lichen sampling and SO_2 measuring sites was highest for Zuid-Holland, the data for this region were analysed first. It is evident from the lichen-distribution map that there is an area where almost all of the species are absent. On the basis of occurrence of species, three zones can be distinguished here:

Zone A: The zone poorest in lichens, with only 4 species:

- *Lecanora conizaeoides* - This species occurs almost everywhere but is sterile in many places in this zone.
- *Buellia punctata* - Occurs rather infrequently in this zone and is absent in the Westland area and around Rotterdam.
- *Lepraria incana* - Occurs almost everywhere in this zone.
- *Lecanora dispersa* - Occurs in some places in this zone.

Zone B: The species of zone A and also *Lecanora expallens.*

Zone C: The species of zones A and B and others as well.

The borders of the zones were drawn (see Fig. 57) such that only the four above-mentioned species occur in zone A, and in zone B only these species plus *Lecanora expallens*. In both zones there are places in which none or only some of the 'allowed' species occur, since the criterion was not the absence but the presence of certain species. *Lecanora dispersa* occurs only sporadically and was not taken into account in the determination of the borders. In zone B 'islands' of zone C occur on which different (more) lichen species than those of zone B, were found at one or more sampling sites. In zone C epiphyte richness increases distinctly with distance from zone B, but the rather small number of sampling sites in this richer part of Zuid-Holland made it difficult to indicate zone borders here.

Comparison of the maps showing the lichen distribution with those showing the SO_2 concentrations at the measuring sites permits the following conclusions:

a) Zones A and B coincide with high sulphur dioxide levels.

b) The agreement between the lichen zones and the SO_2 levels is poorer for the summer or entire year SO_2 levels than for those of the winter period.

c) For the lichen-poor area, the agreement is best for the 50 and 75 percentiles of the winter-period; for the 98 percentile the agreement is appreciably lower.

d) The agreement is best when, for each measuring site, use is made of the value of the winter with the highest SO_2 levels for that site.

e) Zone A corresponds with the area in which the 50 percentile lies above 100 $\mu g/m^3$ in at least one of the two winters for which data are available.

f) Zone B corresponds with the area in which the 75 percentile is higher than 100 $\mu g/m^3$ (or the 50 percentile above 75 $\mu g/m^3$, but here the agreement is slightly lower) in at least one of the two winters for which data are available.

The above-described zones are the easiest to define, but conclusions a-d would still hold if zone borders had been chosen differently; the limiting concentrations would of course be different.

The differences in degree of agreement between the lichen-poor areas and the various percentiles of the SO_2 concentrations (conclusion c) are also evident from Appendices 58-60. These graphs were made by determining the lichen zone in which each of the SO_2 measuring sites fell. In cases where the measuring site was situated (almost) on the border of a lichen zone, the site was assigned to the poorer of the two zones (see Appendix 62). The concentrations used in this graph are the highest measured concentrations (in the category in question) at each site. For the resulting graphs it was then determined, for each percentile, which concentration showed the best agreement with the border between two zones (see Appendix 61a). For instance, with a 50-percentile value of 100 $\mu g/m^3$ as limit there are two sites in zone A with a lower concentration and four sites in zone B with a higher concentration, forming the six overlapping points (sites) in Appendix 61b.

The border between zones A and B corresponds best with a 50 percentile concentration of 100 $\mu g/m^3$ SO_2, that between zones B and C with a 75 percentile concentration of 100 $\mu g/m^3$. The 75 percentile also corresponds (at a concentration of 145 $\mu g/m^3$) reasonably well with the A/B border. The measuring sites with an SO_2 level diverging from this rule are shown in Appendix 62.

Percentile values for summers and entire years were plotted in the same way, again taking the highest value for a given season. The sites under the influence of the coastal effect were not taken into consideration. The values in Appendix 63 indicate the amount of overlapping between zones. It is clear that the differences in SO_2 level between the zones are most distinct in the winter and smallest in the summer (greatest overlapping). The percentiles for an entire year show only slightly more overlapping than those for the winter period, especially for zones A and B. Again for the A/B border (and here also for B/C), both the 50 and 75 percentiles agree well with the epiphyte zones.

It proved possible and easy to draw the zone borders in such way as to hold not only for the lichens but also to include all of the measuring sites with a 50 or 75 percentile higher than 100 $\mu g/m^3$ in zone A or zone B respectively. It is these borders that are shown in Appendices 50, 57, and 64. When using this borders there are no "overlapping" SO_2 sites. Comparison of the borders with the location of the measuring and sampling sites shows where the position of the border was determined not only by the lichen distribution but also by the SO_2 levels at the measuring sites (see also Appendices 56 and 62). With decreasing number of sampling sites the influence of the measuring sites increases and the border is drawn rather more arbitrarily as far as the lichen distribution is concerned. The irregularity of the border increases with increasing density of sites. There is probably a zone of gradual transition, and the irregularity would show a slightly different picture if the location of the sites were slightly different.

10.4 Testing of the conclusions

Conclusions a-f concerning the relationship between sulphur dioxide levels and the distribution of lichens were tested by determining whether they held for the area in the province of Noord-Holland for which SO_2 data were available. In this area the number of sites for both lichens and SO_2 was appreciably smaller, leading to lower accuracy than was possible for the Zuid-Holland area.

The conclusions proved to hold for Noord-Holland as well. Here, too, the 50 and 75 percentiles of the winter SO_2 values gave the best agreement with the lichen distribution. The zone borders could be drawn in the same way as for Zuid-Holland (see Appendices 57 and 64). For Noord-Holland, a concentration of more than 100 $\mu g/m^3$ SO_2 in at least one of the three winters was taken as norm (for both the 50 and the 75 percentiles), since data were available for three winters.

The small number of sites made it difficult to determine the border in many places, but there were no places for which the data were in conflict with the conclusions drawn for Zuid-Holland. Therefore, there is no reason to reject the conclusions.

10.5 Zonation in the part of the area richer in lichens

Examination of the lichen-distribution data for the entire study area showed that zones could also be indicated in the areas richer in lichens. For the determination of these zones, use was made of the conclusion drawn by Brand and Meltzer for their areas (see Brand 1976 and Meltzer 1976). The distinguishing of the zones was not possible on the basis of the occurrence of certain species, and was done according to the numbers of species. We made the following classification: for the frequently occurring tree species the number of epiphytic lichen species present at a sampling site is indicated:

	Zone C	Zone D	Zone E
Ulmus	0 - 11	12,13,14,	> 15
Populus, *Tilia*, *Fraxinus*	0 - 6	7,8,9,	> 10
Salix	0 - 7	8,9	> 10

Zone C was only assigned in cases where other species than those of zone A or B were present.

Only tree species occurring in large numbers were included. The differences in number of lichen species on the various tree species in the separate zones agree well with the classes used for the squares map. The borders were consistently drawn as close as possible to the first sampling sites of the next (=epiphyte-richer) zone (see Appendix 57). The resulting zonation is shown in Appendix 64.

The lines drawn to mark the borders between the zones are, however, rather arbitrary, because the transitions are gradual and there are rather wide differences within zones. It is quite clear, in any case, that relatively epiphyte-poor trees occur within all of the zones.

The relationship between the sulphur dioxide data and the lichen data is less distinct for these richer zones. To obtain an impression of the SO_2 maps and the zone map were compared (see also Appendices 58-61). The best correspondence is given by the 98 percentile, especially for the border between zones C and D (limiting concentration: 150 $\mu g/m^3$ SO_2, 98 percentile). Between zones D and E there is little difference in SO_2

level. It might be possible to draw conclusions for these zones when data for more years were available, since the number of lichens may have been limited by high SO_2 levels occurring incidentally some years earlier.

It is not possible to draw the borders C/D and D/E such that correllation between SO_2 and lichen zones becomes better, as was the case for borders A/B and B/C. When determining the borders on the basis of both lichen sites and SO_2-sites the number of "overlapping" sites stays about the same as in Appendix 61b. (See also Appendix 56).

10.6 The drawing of conclusions concerning SO_2 levels on the basis of lichen zones

Comparison of Appendix 64 with Appendix 55 shows that lichens are most scarce in the large cities, the industrial areas, and the areas where crops are raised commercially in glasshouses. This and the correllation between lichens and the SO_2 levels suggest that the SO_2 emission due to heating (and industrial processes) is the cause of the scarcity of lichens. Because correllation with SO_2 is very strong in this region, conclusions can be reached concerning the SO_2 levels in parts of the area in which no measuring sites are located. It is clear from the zone map (Appendices 57 and 64) that in areas with glasshouses but without measuring sites, the SO_2 level must be high too, for instance the surroundings of the village of Bleiswijk belong to zone A and during or before the survey period the median concentration was probably higher than 100 $\mu g/m^3$. The same holds for the area around Aalsmeer. For the area around Hoorn, where there are also glasshouses, albeit fewer than in the Westland region, it is clear that there are fewer lichens and therefore probably higher SO_2 levels than in the surrounding area.

It is also clear that when only measuring-site data are available (see Appendix 2), different and probably less correct conclusions will be drawn concerning the distribution pattern of sulphur dioxide. Another conclusion is that the method described above can only be applied when the number of sites is very high.

10.7 Discussion

10.7.1 The distribution of the lichens

Very large differences in epiphyte richness were found to occur in the western part of The Netherlands. On the one hand there are areas with onlya few species, and on the other there are large epiphyte-rich areas in the northern part of this region and in some places in the coastal strip. When this difference is described by dividing this area into zones of comparable epiphyte richness it is evident that in the poorest area (zones A and B), all but a very few species have disappeared. The remaining species are evidently very insensitive to sulphur dioxide. Although Lecanora expallens (Appendix 65) is not sensitive to high peak concentrations, it disappears when the median SO_2 value becomes too high.

Zone B includes many islands (13) where other species also occur. All of these islands are formed by old trees standing in a protected place and having a eutrophiated bark; many of these trees are situated near or on farms. The lichens on them are poorly developed and often occur only at the base of the tree or in cracks in the bark. Physcia tenella (21;10, i.e., recorded at 21 sites in 10 islands) occurs on most of the islands, and Buellia canescens (6;5) and Xanthoria parietina (6;5) were also found rather often. The other species observed on the islands were: Parmelia sulcata (4;3), Xanthoria polycarpa (2;2), X. candelaria (1;1), Lecanora chlarotera (2;2), Lecidea limitata (4;3), Catillaria griffithi (1;1), Physcia caesia (1;1), Ph. grisea (1;1), Parmelia physodes (1,1) and Ramalina farinacea (1;1).
The high frequency of Physcia tenella in particular need not be due to limited sensitivity alone but could also be the result of the absence of a strong habitat preference in this species. Although it was found in 20 sites in zone B, other 'extra' species also occur at 14 of them, and of the 6 places where this was not the case, 4 lie very close to places where it was. Thus, Ph. tenella is clearly restricted to places where other species can survive as well, and is therefore a good indicator for places with slightly more species than the numbers in zones A and B.

It is clear that all these species on the islands belong to one group of rather insensitive species (but more sensitive than Lecanora expallens), i.e., the species which are more or less frequent in zone C and have persisted in slightly less polluted places in zone B. For two of the islands it was found from the measuring-site data that these are indeed small areas with lower SO_2 levels: in neither of the two winters did the 75 percentile lie above 100 $\mu g/m^3$.

In zone C a large group of species occurs (more than 20) but the number of species per sample is very small. This is due to the sporadic occurrance of these species, for instance Ramalina farinacea (see Appendix 69). Other species are slightly more frequent, for instance Lecanora subfusca s.l. (this is mainly L. chlarotera) (see Appendix 68). Only a few species (e.g. Ph. tenella and P. sulcata; see Appendix 66,67) have a high frequency in this area, and therefore for the zonation the number of species per sample is a better norm than the species composition. But even on the basis of the number of species the drawing of a border between zones C and D remains rather arbitrary. The transition is gradual: from zone C to zone D the frequency of the species increases and there are more often higher numbers of species per sample. The species composition also changes in the sense that additional species occur. The same holds for the transition from zone D to zone E in even higher degree. Especially in the richer zones the composition of the samples is rather variable due to the occurrence of less common species. As a result, the increased species richness is only well expressed if the numbers of species in several samples were taken together, as was done in the analysis of the squares (see Chapter 5). There too, the number of samples required per squares was found to increase with increasing epiphyte richness, but more than four (only at very great epiphyte abundance) sampling sites are seldom needed. Appendix 70 shows schematically how the number of species in a sample increases with increasing epiphyte richness and how the frequency of samples with a given number of species increases.

The border of a zone is taken where the first sites belonging to a new category appear. With increasing lichen richness there are more trees with steadily increasing number of lichen species. In zone E of this area it is not clear whether trees with an even higher number of species occur. The epiphyte increase seems to be expressed mainly in a higher frequency of epiphyte-rich trees and greater differences in composition between samples. This might be explained on the basis that here air pollution is not the primary limiting factor, but that other factors are also decisive for the number of species. If air pollution by sulphur dioxide is relatively less important here, this would also explain the obscure relationship between the D/E zone border and the SO_2 concentration (but see Chapter 12.5 and 17).

10.7.2 The zone borders

From Appendix 57 it can be seen that the borders cannot always be drawn with the same certainty. For the border between zones A and B in Zuid-Holland there is a reasonable degree of certainty, especially because the number of sites is very high here, and only in the southwestern part of the zone A area around Rotterdam is the line based on very few points. There are almost no suitable trees in this area, and since almost no samples could be taken the border is based mainly on the SO_2 measuring sites (see also Appendix 56). The northeastern part of the A/B border near Amsterdam is based on very few sampling sites, but here there were a considerable number of measuring sites. The border between zones B and C is based entirely on the measuring sites here, and the line is also arbitrary because the SO_2 levels at some sites lie very close to zone C concentrations. This area should be surveyed more intensively.

In Zuid-Holland the border between zones B and C is very irregular and would undoubtedly have shown different irregularities if different sampling sites had been chosen. Thus, the line shows approximately where the border runs. In this border area there is a rather vague transitional zone. Around the islands the line is not always equally distinct either, especially when few other sampling sites lie in the vicinity.

The borders between C/D and D/E are rather clearcut in the northern part of the area, and in the west the same holds for the D/E border. Here, the C/D border is difficult to indicate, especially between Haarlem and 's-Gravenhage, and the line shown in Appendix 57 is a rather rough approximation. The same difficulty was encountered in the eastern part of the area for the D/E border, because the frequency of E sites was low. This was also the case in the southern part of the region: there are sampling sites with many species but the frequency is consistently low.

The C/D and D/E borders are easier to define in the northern part of the region, probably because the epiphyte richness increases here strongly over a short distance. Elsewhere, epiphyte richness also increases near the borders, but there the number of species never becomes as high as in the northern part (see Appendix 70). This can also be seen on the squares map: in the north the squares close to the D/E border fall in class 6, but this is not the case elsewhere.

10.7.3 Correlation of lichen zones with SO_2 concentrations

Comparison of the lichen and the sulphur dioxide data showed that when the 98 percentile lies above a certain value the number of species per sample is smaller because many species occur less frequently or not at all. At these high peak values a large group of species still occur, but in many cases less frequently, and disappear completely when the 75 percentile of the winter period lies higher than 100 $\mu g/m^3$. Under these conditions only a few species remain, and of those Lecanora expallens too disappears if more than half of the days in the winter period have more than 100 $\mu g/m^3$ SO_2.

The correllation is most distinct in the most severely polluted area. Several factors play a role here:

1. for this area the criterion is the complete absence of species, which gives a more distinct border than does increased frequency of several species; and
2. in this area the number of both measuring and sampling sites was much higher than elsewhere.

Nevertheless, there may be said to be a gradual transition here too, which gave rise to the irregular course of the border.

The SO_2 concentrations indicated for each zone are those of the measuring sites in that zone, which are not the same as the concentrations at the sampling sites. The latter will have been much lower, since the lichens often grow in cracks in the bark or at the base of the tree and the trees themselves may stand protected from the prevailing wind. When these factors lead to very favourable conditions, the sampling sites actually belong to a better zone; this might hold for the islands in zone B and 'D sites' in zone C. It is certain, however, that lower SO_2 levels may prevail for the islands at greater distances from the sampling sites, as shown by the measured values at two of the islands.

In all likelihood the 98 percentile varies rather strongly from place to place, especially at lower concentrations, due to such factors as shielding. This too can obscure the relationship between the measured SO_2 levels and the epiphyte levels. Furthermore, it remains possible that in the richer zones the number of species is limited by SO_2 levels in the rather distant past. Certainty on this point would require series of data over a period of many years, and these are unfortunately not available because the measurement of sulphur dioxide began rather recently, particularly in less polluted areas.

It is also undoubtedly of importance that in areas with high SO_2 levels, too, a role is played by various factors (e.g. other pollutants) with a negative influence on lichens, but the strong correlation between the occurence of lichens and the SO_2 levels indicates that in such places the influence of SO_2 is greater than that of other factors. The effect of these other factors becomes more visible where the SO_2 levels are lower. At low SO_2 levels large differences in epiphyte richness can develop locally when other factors vary, for instance when there are places with a group of very old trees standing in a well-protected spot.

11. COMPARISON AND SYNTHESIS OF THE RESULTS OBTAINED WITH THE USE OF SQUARES AND OF SAMPLING SITES

The application of two methods gave rise to the question of the degree to which the results of the analysis on the basis of sampling sites and those obtained on the basis of the squares are in agreement. Comparison of the zone map (Appendix 57) and the squares map (Appendix 12) shows good correspondence. In comparing these maps it must be kept in mind that the class of a square is consistently based on the total number of lichen species on the tree species with the highest epiphyte richness in that square and that two or even three zones often occur in one square. In these cases the epiphyte class assigned to a square is based on the part of the square with the highest epiphyte level. As a result, on the squares map the epiphyte-poor areas are smaller than is actually the case. Details of the zonation can be seen more clearly on the zone map, and with this in mind it is clear that zone A corresponds roughly with class 1, that the B/C zone border lies somewhere in class 2, that class 3 coincides with zone C (i.e., the epiphyte-richest part), and that class 4 corresponds with zone D and class 5 + 6 with zone E.

The squares map is based on the number of species per square. This number is larger than the number of species per sampling site in the same zone (comp. Appendix 6 and Appendix 57). The difference in species number between a sample and a square belonging to the same zone is greater the richer the area is in epiphytes and the more the samples differ from each other in species composition.

According to the classification for squares *Ulmus* has too many species (0-8) in class 1, when this class is identical to zone A (maximally 4 species). An explanation can be suggested for this: the sampling sites with 5-8 species on *Ulmus* in class 1 squares often lie in zone C in squares with a steep gradient. Evidently, rather many species still occur on *Ulmus* at high SO_2 levels. This phenomenon is slightly exaggerated in the classification, possibly due to the following effect associated with the correction applied to the classification. In the correction for *Ulmus*, a comparison was made with the map on which *Ulmus* was omitted, and it is conceivable that on the latter map the number of class 1 squares is too high, because in some squares only *Ulmus* was examined thoroughly and the other tree species were not representative and/or there are many zone C squares where *Ulmus* had many species but the other tree species were very poor in epiphyte species. In both cases class 1 would have been incorrectly assigned on the map for those squares. It would probably have been better to use the combined map including all tree species for comparison in applying the correction, but it was impossible to judge in advance whether a larger error would result from omission or inclusion of the tree species to which the correction was to be applied.

Graphs of the sulphur dioxide concentrations (see Appendices 71 and 72) were made on the basis of the measuring sites in the same way as for the squares (Appendices 46 and 47). The results show that the classes and the zones are also comparable with respect to the correlation with SO_2.

Appendix 73 shows for the 50 and 98 percentiles combined, the zones and classes to which the SO_2 values belong. The oblique line forming the border between zones B and C corresponds about with a 75 percentile of 100 $\mu g/m^3$.

In this graph the 50 percentile is decisive for the epiphyte richness at very low peak values (<120 $\mu g/m^3$ SO_2). Whether this is actually the case is very uncertain, because this segment of the border between zones D and E cannot be drawn with great certainty. Furthermore, the high 50 percentiles never occur in combination with low peak values. Here, the curve would have to be determined experimentally. In this part the line is mainly determined by squares belonging to class 4 where application of the coastal effects makes the SO_2 concentration of the 98 percentile relatively very low.

When this new model (50/98 III) is applied to the 100 squares, the prediction is better than that obtained with 50/98 II because of the diminished number of large errors (comp. Appendix 73 with Appendix 39).

Appendix 73 also shows where the borders lie when the 75 and 98 percentiles are combined (= 75/98 III). The results are less satisfactory (as compared with 50/98 III) for both the SO_2 sites and the squares data. The zones and the classes show more sites or squares with the wrong SO_2 concentration. Nevertheless, this new model too is better for the squares than the old model (comp. Appendix 39). For the rest this model is comparable with the 50/98 III model.

Thus, it may be concluded that when the frequency distribution of the SO_2 levels is known, the epiphyte level can be predicted with a reasonable degree of certainty. The prediction is better the higher the concentration. Especially at lower concentrations, as a result of many factors, the epiphyte levels can be lower than expected (see also section 10.7).

Conversely, it is also possible to read the probable SO_2 level at a known epiphyte richness from Appendix 73, although it holds here too that the uncertainty increases the lower the concentration.

Another factor which can obscure the correlation between the epiphyte richness and sulphur dioxide levels is that low epiphyte richness can be due to the earlier occurence of high sulphur dioxide levels from the effects of which the epiphytes have not yet recovered (see also Chapter 17).

12. RANKING OF SPECIES ACCORDING TO SENSITIVITY

12.1 Introduction

Almost all surveys have made use of or have drawn conclusions concerning the relative sensitivity of the lichen species under study. In many cases this sensitivity was derived from the distance from the source of pollution, where the species was absent (Barkman 1958; Skye 1968; Hawksworth and Rose 1970; Jürging 1975) or from the mean number of companion species (LeBlanc and de Sloover 1970; Horntvedt 1975), or from a combinatio of experimental and survey results (Jürging 1975). (See also section 3.3)

The objection to the first of these methods is that exact determination of the distance is very difficult. This difficulty arises from the fact that the frequency is very low where the species starts to occur (can occur). The exact sequence in which the species are found is highly dependent on chance factors such as the presence of suitable habitats (which are not the same for all species) and the related general rarity (i.e., not determined by air pollution) of the species. This problem is reflected in the data of the detailed survey done in the western part of The Netherlands, i.e., for the species occurring first in Zone C. The frequent absence of a single, clearly defined source of pollution also limits the applicability of this method.

The use of the number of companion species imposes severe restrictions, since only vegetations of the same association occurring in comparable places can be compared (see also section 3.3; where this sensitivity value is used for an I.A.P. determination, the number of species was included twice in the calculation).

The combined method applied by Jürging is rather coarse (only 3 classes) and employs an arbitrarily determined weighted average derived from several different methods of sensitivity determination.

12.2 Method used to determine the sensitivity ranking

As a result of the WHEN project, we now have the disposal of a distribution map for each of the surveyed species. Furthermore, the epiphyte richness of each square was determined. Thus, we could determine for each species whether it occurred proportionally in all epiphyte classes or was to some degree limited to epiphyte-rich squares. (The objections to de Sloover's method do not hold here, because we were dealing with a standardized epiphyte richness per square). The more a species is limited to epiphyte rich squares, the more sensitive the species to the factor(s) responsible for the epiphytic impoverishment. This reasoning was applied for the formulation of a sensitivity ranking for the species found in the VHEN survey. The method used for this purpose can be explained best on the basis of a number of examples.

For comparison of the species it should be possible to compare the 'presence percentages' in each class (for instance, Parmelia physodes occurs in 81.2% of all squares of class 6), but these percentages are strongly influenced by the rarity of a species. A better solution is to determine for each species the cumulative percentages for the various classes, putting the total number of squares in which a species was found at 100%; for instance, from all squares with P. physodes we have 0.6% in class 1, 8.5% in class 1 + 2, 24.8% in class 1 + 2 + 3 and 100% in class 1-6 (see Appendix 74). These percentages can then be compared with the cumulative percentages of all squares. If a species occurs proportionally over all of the classes, the cumulative frequency distribution of the species is the same as that of all of the squares. When this is plotted graphically, the result approximates a straight line under 45° (see Appendix 75).

This method offers two distinct advantages:

a. The results are independent of the rarity of the species i.e., of the number of squares in which the species was found.

b. The results are independent of the classification method. The only assumption made is that the squares of class 2 are richer in epiphytes than those of class 1, and so on. There is no need to assume that, for instance, squares of class 2 are twice as rich as those of class 1. In addition, the outcome is virtually independent of the total number of squares of a given class (i.e., of the total distribution of all squares over the classes). For, the graph can be seen as an approximation of the outcome if the exact rank of all squares had been known. With different numbers of squares per class, the points on the graph would have been different (i.e., with different values on the horizontal axis) but the shape of the curve would not be different. A familiar application of this type of graph is the Lorenz curve for the comparison of income distributions (Pen 1970). The sensitivity curves of a number of species are shown in Appendices 75 and 76.

Differences in the shape of the curves can be seen most clearly in the middle segment, from the point between classes 2/3 to the 5/6 point. From the latter on, all lines converge toward 100%, and before the former toward 0%. Only very insensitive species occur in class 1 at a rate of 1% or more. For the determination of the ranking of sensitivity, therefore, only this middle portion of the curves was used; in this segment the differences are also greatest. When a species is very seldom found, the reliability of this method is lower. For example, when a species was found in 50 squares, one square more or one less amounts to a difference of 2%.

For all species found in 25 or more squares, a curve was plotted in this way. The corresponding percentages are shown in the table in Appendix 74. For the species that were not surveyed in all squares (see Appendices 3 and 13), the relevant squares had to be omitted on the horizontal axis as well. As a result of this omission, however, the class limits on the horizontal axis did not shift by more than 1% (see Appendix 74), which is a negligible difference. The number of unsurveyed squares per class is invariably small, and therefore it is not to be expected that the absence of data led to appreciable differences.

12.3 Results

A high proportion of the species could be easily assigned a place in a sensitivity rank order. These species have curves with a comparable shape and little or no intersection of the limits. On the basis of the curves, eight groups of approximately equal sensitivity can be distinguished. The limits of these groups (sensitivity classes) were chosen such that the smallest possible number of curves intersects the limits.

Appendix 77 shows schematically the sensitivity classes to which the species were assigned. Many species can be assigned directly and have curves that do not intersect. Another group of species sometimes have curves lying very close together and intersecting each other but not the class limits. A third group has curves diverging from most of the others in shape and therefore intersecting the class limits. These species show a behaviour deviating from that of most of the species: many of them occur relatively often in the epiphyte-poor squares (class 2). Thus, these species, exemplified by *Pertusaria amara*, may be slightly less sensitive to high concentrations of air pollutants than other species with a comparable curve occurring in epiphyte richer areas. It is, however, also possible that such differences (which often do not amount to more than a few per cent) arise from chance factors, for instance because the species was often found in the epiphyte-rich part of squares. The reverse is also seen: species occurring in relativity few epiphyte-poor squares (e.g. Catillaria griffithii). Nevertheless, the deviations are always rather small and it is usually clearly evident to which sensitivity group one of these species belongs.

A separate group is formed by Cetraria glauca, Lecidea granulosa, Parmelia saxatilis, and Parmelia tubulosa. These species occur relatively often in class 3. It is noteworthy that these are all species for which the free-standing standard trees of the survey do not offer an optimal habitat. This might explain a divergent sensitivity curve, but it is also possible that these species were always found in the most protected sampling sites and that the latter are not representative of the epiphyte class assigned to the square, although this does not explain the surplus in class 3.

In Appendix 78, the 'presence' curves of a number of representative species are shown, because visual interpretation of those curves is somewhat easier than for the material in Appendices 74-76.

For very common species it remains possible that the picture given by the above-described method is distorted. It is conceivable, namely, that for them the frequency can increase although the species occurs in almost all squares. This increased frequency would then occur only within the squares. When two species have highly divergent frequency courses within squares, this may cause inexact differences in sensitivity. However, only a small number of species are that common, and their sensitivity rank was confirmed by the analysis of the sampling in the western part of the country. Parmelia sulcata is also more sensitive than Physcia tenella when the above-described method is applied with use of the zones of this region (see Chapter 10) as unit instead of the classes. Thus, for the very common species, too, the cumulative frequency distributions give a good reflection of the sensitivity ranking.

The ranking can be less reliable for the rather uncommon species, since when a species is found in 25 squares one observation makes a difference of 4%. This limitation holds, for example, for Arthonia radiata and A. spadicea.

The sensitivity of a given species is partially dependent on the surroundings and the substrate. As discussed in Chapter 13, the sensitivity on different substrates can be rather divergent. The sensitivity indicated by the total distribution of the species is an average that is dependent on the variation in sensitivity within the individual species and on the ratio of the phorophytes in the distribution area of that species. In most cases the total sensitivity will lie close to the lowest sensitivity.

There are few species that are entirely absent in class 1, but most of those present do not exceed 1% (Appendix 74). It is not possible to indicate precisely the epiphyte class in which the various species appear; the frequency increases gradually, and there is no distinct absolute limit The 'limit' below which say 10% of the species occur can of course be indicated (for sensitivity class A this limit lies in epiphyte class 2, for B-E in class 3, and for F, G and H in class 4; see Appendix 76). Such limits cannot, however, be applied in the field.

The gradual frequency increase is not surprising. There are very many variables that influence the occurrence or absence of a species and interfere with the influence of air pollution. This, combined with the variation in sensitivity within a species, means that the limits are not absolute. The method applied also leads to some spread, e.g. due to differences within squares.

12.4 Comparison with the results of other authors

In view of the foregoing we, unlike Hawksworth and Rose, cannot indicate the sulphur dioxide concentration at which certain species disappear, although we can say that zone limit A/B holds for Lecanora expallens and zone limit B/C holds for many other species (see Chapter 10). It is, however, possible to compare the sensitivity ranking derived from the present study with those given by Hawksworth and Rose (1970) and Barkman (1958 and Appendix 3).

For most of the species the three rank orders are roughly in agreement, although there is considerable spread:

WHEN	Barkman	Hawksworth & Rose	exceptions
A	1-3	2-3	Physcia tenella (B:5; H&R:5); Cetraria glauca (B:10; H&R:5)
B	2-4	3-5	Parmelia sulcata (B:6)
C	3-7	4-6	
D	5-8	4-6	
E	±8(6-9)	6	Evernia prunastri (B:8)
F	7-10	6-7	Pertusaria albescens (B:6); P. amara (H&R:5); Parmelia saxatilis (H&R:4)
G		(7-8)	
H	8-11	(8)-9	Parmelia revoluta (H&R:6); P. caperata (H&R:6)

Physcia tenella is a striking exception. Not only Barkman but also Hawksworth and Rose consider this species more sensitive than the present results indicate. In addition, Skye (1968) and Jürging (1975) concluded that this species is just as sensitive as i.a. Evernia prunastri or more sensitive, respectively. As discussed in Chapter 13, the sensitivity of Ph. tenella varied widely on different substrates. In this case the sensitivity is perhaps determined to a high degree by such conditions as the most frequently occurring on surveyed phorophyte species, possibly combined with differences in sensitivity between ecotypes or populations. Cetraria glauca is considered more sensitive by Barkman and Hawksworth and Rose than was found in the WHEN survey. The true sensitivity of this species (and e.g. Parmelia saxatilis) is not clear, since these species showed divergence with respect to both ecology and sensitivity in the present study (see section 12.3).

A striking finding from the comparison with the sensitivity ranking of other authors is that the extremes of these series are in reasonable agreement, whereas for the intermediate group of the moderately sensitive species there is a very large spread. The reason for this is probably that such differences are more difficult to demonstrate with a more qualitative method than with the quantitative method we applied. It is also striking that most of the species in Hawksworth and Rose's zones 8-10 are mainly species requiring a different microclimate than is needed by most of the species of the poorer zones. Many of these species have already disappeared in The Netherlands (e.g. Lobarias and various Usneas) or have become very rare. But it is not clear whether this severe reduction in both Britain and The Netherlands is caused entirely by air pollution or (also) by other factors.

12.5 The occurrence of species of sensitivity class H in areas belonging to epiphyte class 6

Class H comprises species whose occurrence is almost completely limited to squares belonging to epiphyte class 6. But class-H species do not occur in all class-6 squares. In The Netherlands there are two large areas belonging to class 6 but lacking these most sensitive species; for the remainder class 6 is invariably accompanied by species belonging to class H. These exceptional areas form part of the province of Noord-Holland and the region west of Utrecht (compare App. 79 with App. 12). In these areas a large number of species occur but not the most sensitive. These species did occur there at one time (around 1950), and have therefore disappeared since 1950 (see also Chapter 17).

It is striking that the class-6 areas without H species comprise the class-6 areas situated closest to large epiphyte-poor areas. This strongly suggests that the most sensitive species were once present but have now disappeared from these areas.

Although it proved difficult to correlate the limit between epiphyte classes 5 and 6 with SO_2 levels, this was probably due to the lack of sufficient data. All of the evidence points to the probability that peak values (98 percentiles) play an important role here but also that measurements must be made at many stations over a very long period to demonstrate this. All this makes it extremely likely that over a long series of years the incidental occurrence of high concentrations of SO_2 (originating from the severely polluted adjacent region) was (one) cause of the disappearance of the sensitive species. (For the correlation of SO_2 with the occurrence of sensitive species in relation to the problem of indicator value, see also Chapter 17).

H species also hardly occur in the Gaasterland region (see northern area on map sheet 15) or in the Noordoost Polder. For the latter area this could be explained without reference to air pollution, since this land was reclaimed relatively recently (the first trees were planted around 1946) and species which do not spread easily might not have become established there yet. This is supported by the indication that the species richness is much higher on the east side (ie.e. the inland side) of the Noordoost Polder than on the west side, which would be expected if the lichens colonize this area slowly from the eastern border. This phenomenon has also been observed, although less distinctly, in the Wieringermeer Polder. The distribution over the new polders is indicative for the dispersion rate of the various species.

Another factor which could be responsible for the absence of certain species in the Noordoost Polder is that Populus is almost the only available phorophyte in this area.

A similar (but less distinctly expressed) phenomenon was found for the species in sensitivity group G. These species occur everywhere in class 5 (and 6) squares except in the northeastern parts of the provinces of Brabant (map sheets 45 and 46) and Groningen (see southeastern part of map sheets 7 and 8) (see also Appendices 79, 80, and 12).

Thus, on the basis of the distribution of the lichens it is possible to differentiate further within the epiphyte classes distinguished on the basis of number of species.

13. THE INFLUENCE OF THE PHOROPHYTE

13.1 Geographic differences

Comparison of Appendices 22-29 shows that the epiphyte-rich and -poor areas coincide to a high degree for the various tree species. In Appendices 80 and 81 it can be seen in which squares each of the eight tree species included in the study were the richest in epiphytes as compared with the other phorophytes in the square. In adjacent squares this is often the same species of tree.

For example, in the northwestern part of the country *Ulmus* is the porophyte richest in epiphytes. *Salix* is the richest phorophyte in the areas of low-moor peat and clayey soils over peat, and *Populus* in areas where many of this species have been planted (e.g. Brabant and the Noordoost Polder). On the sandy soils north of the large rivers, *Quercus* is the richest in epiphytes; in the eastern part of Brabant *Tilia*; in Zuid-Limburg and other places *Fraxinus*. The very low epiphyte richness of *Quercus* in Brabant (as compared with e.g. *Tilia*) is striking.

The cause of this phenomenon is not entirely clear. In general it can be said that the trees have the most epiphytes where they occur in the greatest numbers. The following factors may exert an influence (see also Barkman 1958):

a. Greater phorophyte density may promote the distribution of lichen species adapted to these tree species, especially for lichen species with a limited capacity for dispersal.
b. Differences between areas in the places in which the trees are situated.
c. Climate effects in combination with characteristics of the tree, e.g. 'compensation' for drought by large water capacity of the bark.
d. Geographic variation in the characteristics of the phorophyte (e.g. the structure and chemical composition of the bark), which perhaps exist and could be caused, for instance, by differences in the type of soil.
e. Difference between areas in the varieties of planted tree species.
f. Differences in tree shape between areas e.g. due to heading.

Of these six factors, the first three are the most likely to have an influence.

The above mentioned differences between tree species emphasize the need for investigation of a number of these species, especially in studies covering a large area with great variety in this respect.

13.2 Species richness

Appendix 82 shows classification T_4 for the eight phorophyte species in graphical form. As could be expected, the number of species to which class 6 was assigned shows strong correlation with the total number of lichen species found on that tree species (correlation coefficient $r=0.87$, $p< 0.5$, for comparison with the total number of species in the WHEN survey). In the present study more species were found on *Tilia* than were reported by Barkman; and the new number also fits better with the classification T_4 used for this species. The difference in these results may be due to the fact that Barkman's survey was not very intense just in the region where *Tilia* is the richest in epiphytes (eastern Brabant and central and northern Limburg) (comp. App. 82 with Barkman 1958, map I, page 2). The differences in epiphyte richness in unpolluted air are striking and for the most part unexplained (see Barkman 1958). It is clear, however, that the occurrence of a lichen species is related to the characteristics of the phorophyte, but in many cases it is not evident why many more lichen species occur on one tree species than on the others.

The decrease in epiphyte richness also differs for the various tree species. For instance, the difference between *Ulmus* and *Quercus* is very striking. When an area becomes severely polluted, the number of lichen species drops more strongly for *Quercus* than for *Ulmus*, where it remains rather high for a long time (up to zone B; see also Chapters 10 and 11). In view of the 'normal' number of lichen species on *Quercus*, the decrease of this tree species is also stronger than for most of the other tree species. We can rank the tree species roughly by making use of the maximum number of species in class 1 as percentage of the maximum number of species in class 5, as follows:

Species:	*Ulmus*	*Salix*	*Tilia*	*Populus*	*Acer*	*Juglans*	*Quercus*	*Fraxinus*
%	36.4	26.7	20.0	18.8	16.7	15.4	14.3	12.5
pH	<------- neutral to weakly acid -------->						acid	neutral
bark structure	rough	rough	rough	smooth-rough	smooth	smooth	rough	smooth

(pH values and bark structure taken from Barkman 1958, Skye 1968, and Grodzynka 1970).

Thus, for tree species with a neutral rough bark the epiphyte richness decreases more slowly under increasing pollution than is the case for a species with an acid bark, like Quercus. Species with a neutral smooth bark occupy an intermediate position, and Fraxinus deviates from the above rule. The following factors will play a role here:

a. It takes more time for a neutral bark to become too acid for all epiphytes.
b. More dust accumulates on a rough bark, which could compensate for the injurious effect of SO_2 (e.g. by elevation of the pH).
c. SO_2 is more injurious in an acid than in a neutral milieu.
d. The SO_2 concentration is probably lower in the cracks of rough bark than on a smooth bark.

Many authors stress the effect of the pH of the substrate on the disappearence of species (e.g. Gilbert 1968, 1970; Kunze 1974).

13.3 Sensitivity

Variation in the sensitivity of a species according to the substrate is mentioned several times in the literature (e.g. Baddeley et al. 1973; Kunze 1974), and is usually discussed in relation to the pH of the bark, because the effect of SO_2 (usually applied in experimental work in the form of aqueous SO_2 or $Na_2S_2O_5$ solutions; see e.g. Puckett et al. 1973; Türk et al. 1974, 1975) is stronger at low than at high pH values.

In the present study differences in sensitivity on different substrates were investigated for a few lichen species. For this purpose we determined in which of the squares where a given phorophyte was surveyed, a given lichen was found. The sensitivity of the lichen on different tree species can be compared by use of the method described in Chapter 12. This was done for three species of lichen, i.e., Physcia tenella, Evernia prunastri, and Parmelia physodes. One of these is a neutrophytic species (Ph. tenella), and the other two are moderately acidophytic. The sensitivity curves of these species can be found in Appendix 83. The tree species chosen were Quercus, Ulmus, and Populus, the first two representing the extremes in the series discussed in the preceding section, and the last an intermediate species.

Physcia tenella proved to be least sensitive on Ulmus (the total curve being virtually the same as the curve for this species on Ulmus), and very sensitive (especially in the poorest epiphyte classes) on

Quercus, *Parmelia physodes* and *Evernia* show the reverse picture: most sensitive on *Ulmus* and least sensitive on *Quercus*. Thus, a lichen species seems to be least sensitive on the tree species on which it occurs most frequently (i.e., in its optimal habitat). Tabulation gives the following picture (the values giving the percentage of class-6 squares with the tree species, carrying the lichens; the boxed values indicate the tree species on which the lichen species is least sensitive):

	Quercus	*Populus*	*Ulmus*
Physcia tenella	52.4	86.4	[91.7]
Evernia prunastri	[85.8]	-	43.8
Parmelia physodes	[85.8]	52.4	27.8

It is clear from this tabulation that at least for these species the sensitivity is more strongly determined by the degree of optimality (for the lichen species) of the phorophyte than by the pH of the phorophyte.

Quercus is a tree with a considerably more acid bark than *Ulmus*, but *Evernia* and *Parmelia physodes*, also in the epiphyte-poorest classes, are the least sensitive on *Quercus*. The extremely divergent sensitivity of *Physcia tenella* on *Quercus* might be explained by a combination of characteristics less optimal for *Ph. tenella* together with the unfavourable effect of a low pH.

It would undoubtedly prove worthwhile in the future to pay more attention to factors other than the pH in studies concerning the relationship between phorophytes and lichens as well as in (experimental) studies on the reaction of lichens to a number of other factors.

In any case, the differences demonstrated in the present investigation show clearly that the total sensitivity, as calculated in Chapter 12, is dependent on the frequencies of the phorophytes on which the lichen species occurs. For the three species in question, the total sensitivity is roughly the same or only slightly greater than that of the tree species on which the lichen species is the least sensitive. But when only for instance *Quercus* is surveyed, the sensitivity of *Physcia tenella* is appreciably greater than for *Ulmus* alone or together with *Quercus*. This might partially explain the different sensitivity found by various authors. For instance, it is easy to understand that Jürging (1975) concluded that *Ph. tenella* is a sensitive species (to SO_2), since he investigated only *Quercus*.

The foregoing could also hold for other factors besides the phorophyte. When, for instance, a free-standing tree does not provide an optimal habitat for a given lichen species, the sensitivity found in the WHEN study is probably divergent from the sensitivity that would have been found if the optimal habitat of this species had been included in the investigation. For species occurring optimally in protected spots, the results are influenced by the fact that in such places the air pollution is usually lower than is the case in exposed spots in the immediate vicinity.

14. DISTRIBUTION MAPS

14.1 Introduction

For each of the species of lichens observed during the WHEN project a distribution map was made on the basis of the 5 x 5 km grid. To determine whether the distribution of a species has changed in the course of time, it is necessary to have the disposal of a sufficient number of reliable distribution data collected in the past. The old data made use of here were collected by Barkman. These data concern both his own observations made in his survey of epiphyte vegetations (some of them unpublished field observations, and some published in 1958) and the material he extracted from the literature and the most important Dutch Herbaria (see Barkman 1958, page 246). In a few cases the records carry indications showing that they date from before or after 1925.

Unfortunately, the distribution data collected before the WHEN investigation are less complete, and this makes it difficult to draw conclusions about changes in the distribution of the less common species. Only very marked changes can be directly seen by comparison of the old and the new data. This holds for such species (type IV) as _Anaptychia ciliaris_ (App. 85), _Parmelia perlata_ (App. 95), and _Physcia pulverulenta_ (App. 102). The least of these species, for instance, was found frequently in the central part of The Netherlands in the past and no longer occurs here.

But when the picture is less clear than for these species, it does not necessarily hold that the distribution of a given species has undergone no change. There are many species which occur rather generally but show distinct 'holes' in their distribution. Although there is no direct evidence that in these cases the species occurred formerly where it is now absent, it may be assumed to have disappeared in such 'holes', and in all likelihood due to air pollution. This may be assumed because their absence occurs in places which are poor in epiphytes. And, as we have seen in the preceding chapters, SO_2 pollution is strongly correlated with this epiphyte impoverishment. Examples of species (type II) with this kind of distribution pattern are provided by _Ramalina farinacea_ (App. 103), _Xanthoria parietina_ (App. 104), and _Parmelia physodes_ (App. 96).

But when a species is less generally common it is less easy to see whether it is absent especially in epiphyte-poor squares and thus that its scarcity is probably determined (to a high degree) by air pollution.

Examples of species that are frequently absent in epiphyte-poor squares (type III) are Pertusaria albescens (App. 99) and Cetraria chlorophylla (App. 88). Species which are often absent but less restricted than the above species to epiphyte-poor squares (type V) include Lecanora dispersa and L. muralis (App. 90 and 91, respectively).

Chapter 12 indicates how restriction to epiphyte-poor squares can be compared for rather common species. The very sensitive species (sensitivity class H) belong to type IV; type II comprises moderately sensitive (B-E) very common species; type III very sensitive species (sensitivity class F, G); and type V the species which are rare but moderately sensitive or insensitive in relation to their rarity. The very insensitive species belong to type I (see App. 89).

Concerning the very rare species for which it cannot be inferred from the old data that they were once more common, no conclusion can be drawn about sensitivity. It is almost always difficult to say anything about changes in the distribution of such species.

It is more difficult to distinguish an increase of a species than a decrease. The WHEN survey is much more complete for most species than the old data, and therefore the earlier absence or lower frequency of records of species does not necessarily reflect real absence or scarcity in the past,

Thus, it is to a certain extent possible to distinguish between rarity due to air pollution and rarity due to other factors. For species which are infrequent in relation to the sensitivity class to which they belong, some factor other than air pollution must be responsible for the limited occurrence (see Appendix 84). Another striking point to be noted in Appendix 84 is that in the most sensitive classes (E-H) even the most common species of the class occur less frequently than would theoretically be possible. The theoretically possible number is calculated on the basis of the assumption that the species can occur in all squares belonging to class 6. This assumption is of course the less likely to hold, the greater the sensitivity (since the sensitivity is expressed in declining frequency with declining epiphyte richness and not by a sudden transition between normal occurence and total disappearence). But when sensitive species are more selective for other factors as well (i.e., when sensitive species are generally more stenoecious than intensitive species) the same phenomenon will be observed, because under these conditions other factors in addition to air pollution limit their occurrence.

It is also noteworthy that it is just the species whose sensitivity class is relatively uncertain which are relatively rare. This could be due, for instance, to the fact that free-standing trees are not the optimal habitat for these species. Other factors besides air pollution (e.g. the presence of a forest in the vicinity, or a relatively protected position of the tree) are then extremely important for the occurrence or absence of a species at the sampling sites. These factors would thus cause both the relative rarity and the divergent sensitivity curve (see also Chapter 12 and the location of e.g. Cetraria glauca and Parmelia saxatilis in App. 84). This assumption seems justified, because the species with a 'low' place in App. 89 include many park and forest species.

14.2 Examples

In this section a number of examples of each type of species will be treated. The distribution maps of these species are given in Appendices 85 - 104. The distribution maps of Buellia canescens, Evernia prunastri, Lecidea scalaris, Parmelia caperata, P. furfuracea, Pertusaria amara, Physcia orbicularis, Ramalina fastigiata, R. fraxinea, and Usnea spp. will be found in de Wit (1976). Here, as everywhere in this publication, nomenclature is according to James (1965). Phytogeographic districts are according to Barkman (1958). Only epiphytic records are indicated on the maps and only very few records are from other than standard WHEN-sampling sites.

Anaptychia ciliaris (App. 85)

This is one of the species that have undergone a severe decline. During the WHEN investigation it was found (sterile) in only three places (in the coastal dunes on old Ulmus and Populus trees in protected spots). In the time covered by Barkman (1958) the decline was not yet distinct from the number of records but the species had already disappeared from cities and was seldom fertile (in 15% of the sites as against 78% before 1925; Barkman 1969). According to Hawksworth et al. (1973), this species has an eastern distribution in the British Isles and is becoming very rare in western counties; in large regions it has already disappeared. The comparability of the changes in the distribution of Anaptychia and the genus Usnea mentioned by Hawksworth et al. (1973) does not hold for The Netherlands, where Usnea (de Wit 1976) has declined less than Anaptychia.

In addition to air pollution, the cutting down of very old trees (along roads) may play a role in the disappearance of this species (Barkman 1958), particularly since the decline in fertility began, because these two factors in combination greatly increase the dispersal difficulties.

Anaptychia shows the same phenomenon as other species which are becoming very rare, i.e. withdrawal toward the coastal strip.

Buellia griseovirens (App. 86)

The occurrence of this species is strongly restricted to epiphyte-rich squares, but it is common in epiphyte-rich areas. This permits the conclusion that it is a sensitive species. It is not clear for this species, however, whether the restriction is due to disappearence from epiphyte-poor squares or whether it never occurred in those places. According to Barkman (1958), the species was found for the first time during his study (5 times). This conclusion and the very great difference between these five observations and the 229 made in the present study, suggest that B. griseovirens may be a recent immigrant whose distribution is restricted by air pollution.

Buellia punctata (App. 87)

This species is very common and occurs almost everywhere; it was found to be absent only in very epiphyte-poor squares. Its absence in a few other squares is probably to be ascribed to its having been overlooked during the survey.

Cetraria chlorophylla (App. 88)

This species occurs in the areas with oligotrophic soils. It is rather rare (65 squares), even in view of the rather marked sensitivity (class F). Other factors besides air pollution therefore probably limit the occurrence of this species severely. The rarity also makes it difficult to draw reliable conclusions concerning changes in the distribution, but it seems likely that this species has disappeared from Twenthe, northern-Limburg, and the surroundings of the city of Arnhem.

Lecanora carpinea (App. 89)

This species is rather common (423 squares) and is slightly less sensitive than Cetraria chlorophylla. Air pollution is one of the most important limiting factors for the occurrence of this species. Lecanora carpinea was found on all eight tree species included in the survey and also on old, rough-barked specimens, which is in conflict with Barkman's

(1958) statement that it occurs 'invariably on smooth bark' but in agreement with his report that he found it on rough bark (Barkman 1958, Table XXXVI). It is noteworthy that there are no pre-1955 records of its occurrence in the area where the species is now found most often. This is probably a matter of chance, however, in view of the small number of places in this area for which lichen data collected before 1955 are known.

Lecanora dispersa (App. 90)

This species occurs mainly in the western part of the country and in the 'Southeastern Sanddistrict'. On the basis of the marked insensitivity and the proportionally low number of squares (339) in which the species was found, its distribution must be restricted by other factors besides air pollution, but these factors are not knwon. This species has probably been overlooked fairly often and occurs frequently at non-standard sampling sites. The limited sensitivity of Lec. dispersa to air pollution is also reported by other authors (see e.g. Barkman 1958).

Lecanora muralis (App. 91)

The distribution of this species is very remarkable: it occurs almost exclusively in Limburg and eastern Brabant. Only two epiphytic observations dating from before 1955 are known, and Barkman (1958) calls Lec. muralis an epilithic species sometimes found 'at the base of dust-impregnated trunks'. It is very probable that the present distribution is the result of a marked extension (as epiphyte) in recent decades. No distinct cause can be suggested for this change, however. A possible cause could be the increased surfacing of roads with asphalt and the resulting change in the composition of the road dust, but this does not explain why the phenomenon is (still) restricted to the southeastern part of The Netherlands. Another possibility is that the change is due to the appearence of a mutant which has spread from a point in this region. It is noteworthy that in 1963 Barkman reported no fewer than seven epiphytic sites in Belgian Limburg, all in the northern part of that province, i.e., rather close to the Netherlands Limburg and the eastern part of Noord-Brabant.

Lecidea limitata (App. 92)

This is a moderately sensitive species (class D) and occurs mainly up to within 100 km of the coast. Besides air pollution, the distribution of this species if therefore probably defined by the climate.

Lecidea symmicta (App. 93)

This species occurs mainly on oligotrophic soils and in the Noordoost Polder. It occurs very frequently at places with conditions divergent from those on the standard WHEN sampling sites. This might explain the ill-defined sensitivity (moderate to fairly sensitive; roughly class E) and its relative rarity.

Parmelia acetabulum (App. 94 a and b)

This is a very common species (733 squares) occurring throughout the country but absent in very epiphyte-poor squares (sensitivity class D). There are no pre-1955 records for the Veluwe region, the eastern part of the province of Overijssel, the soutern part of Drenthe, and Limburg. It is not certain whether this absence reflects a real distribution pattern or is due to insufficient sampling.

It is noteworthy that this species is virtually absent in the Noordoost Polder (occurring only on the former island of Schokland), which is indicative of a very limited capacity for dispersal.

Parmelia perlata (App. 95)

This was one of the rapidly declining species as early as around 1950 (Barkman 1958). Before that time it had been found in 34 places, around 1950 it was observed in 13 places, and in the present study in 11 places. Like Anaptychia ciliaris, this species has withdrawn to the coastal dunes, but the former species only showed this trend after about 1955, when P. perlata was already restricted to the dune area. All of the inland sites recorded after 1925 date from before Barkman's study. The number of records of this species has shown hardly any further decline during the last twenty years, but this may be deceptive, because the coastal strip (dune woods, etc.) was very thoroughly investigated recently by M. Brand. Most of the recent records concerning this species derive from his study and not from the WHEN project. A distinct indication for a decline is provided by the finding that, despite the thorough survey, the species was not found recently in the coastal dunes between Hoek van Holland and Beverwijk.

Hawksworth et al. (1973) also report a strong decline for P. perlata, similar to that of Anaptychia ciliaris but to a much greater extent than the latter; and 'in England north of the Thames it survives only in a narrow zone near the East Anglian coast'. These authors think that the decline began very early in the Midlands, which is consistent with the finding that here too its reduction started much earlier than that of *An. ciliaris*.

Parmelia physodes (App. 96)

This species is very common throughout The Netherlands (800 squares) and is absent almost only in epiphyte-poor squares (class B, type II).

Parmelia revoluta (App. 97)

This is a rare, very sensitive species in The Netherlands (55 squares, class H) which is now absent in rather many places in which it formerly occurred, e.g. on the coast in Zuid- and Noord-Holland, the western part of the province of Utrecht, and in the eastern part of the country in the region of the large rivers.

Parmelia sulcata (App. 98)

A very insensitive and very common species absent only in squares with very few epiphytes (1219 squares, class B). There are a few pre-1955 records for the vicinity of Amsterdam and The Hague, where it no longer occurs.

Pertusaria albescens (App. 99)

This is a rare species (210 squares) occurring mainly in epiphyte-rich squares (sensitivity class F). Recently, it has also been found frequently in the Drentian and Southeastern Sand district, in contrast to the pre-1955 situation, but its absence there in the old records could be due to a combination of the rarity of the species and insufficient sampling.

Pertusaria coccodes (App. 100)

This species is both very rare (58 squares) and very sensitive (class H), and is found mainly on the oligotrophic soils north of the large rivers. The pre-1955 data are too scanty to permit comparison with the present situation. On the basis of the distribution map for *Pertusaria coccodes* in Britain (James and Rose 1974), the species seems to be less sensitive than *Parmelia caperata* there.

Physcia aipolia (App. 101)

One of the extremely reduced species in The Netherlands. Before 1955 it occurred everywhere on eutrophic soils, but is now restricted to the coastal zone. Records are known for 69 squares before 1955; recently, it was found in only 19 squares. Because of the small number of squares the sensitivity was not calculated, but it must certainly be of the same order as that of the species in class H.

Physcia pulverulenta (App. 102)

This species too is declining strongly. It was found in 70 squares in the present study and in 104 before 1955. Even more important, however, is its disappearance from the central part of the country. For a species belonging to sensitivity class F, it may be considered extra rare. This forms an indication that some factor in addition to air pollution limits the distribution of this species. Interference with this factor may be responsible for the slightly divergent course of the sensitivity curve of Ph. pulverulenta.

Ramalina farinacea (App. 103)

This is a very common species (776 squares), and there are no areas where it was recorded before 1955 but is now absent. It is, however, almost entirely absent in squares belonging to epiphyte classes 1 and 2 and even in epiphyte-rich squares its vitality is much lower than it was several decades ago. In Great Britain , too, R. farinacea is one of the least sensitive of the Ramalina group (Hawksworth et al. 1973).

Xanthoria parietina (App. 104)

This species has a distribution roughly comparable to that of R. farinacea, but is slightly less sensitive than the latter. It is noteworthy, however, that compared with R. farinacea and P. acetabulum, it occurs much less frequently on the oligotrophic soils north of the large rivers. This might be due to the fact that in this region the oak was one of the most frequently surveyed trees and X. parietina occurs only on strongly eutrophiated oaks.

14.3 Other species

In addition to the 59 species found in at least 25 squares, as listed in Appendix 74, many other species were also found in fewer squares. The number of the latter cannot be given exactly, because some of them could not be determined with certainty, but the total number of species occurring at the WHEN sampling sites is at least 120. The 16 species found in 10-25 squares are listed in Appendix 105. All of the remaining species were found less frequently.

For many of the less frequently found species the number of squares is not representative of the true rarity. This group comprises species which are often difficult to recognize and are easily overlooked or confused with other species (see also App. 4). They are also likely to be species which occur mainly in other places than free-standing trees (see also section 14.1).

There are of course also many other species which are found in The Netherlands but not on the free-standing trees included in the WHEN survey. The total number of recently observed species is at least 150. In addition, there are more than 61 species which were observed with certainty in the past but not recently. This latter number is not really certain either, because: a) some old records concern erroneous determinations; b) there is still undetermined material in herbaria; c) recent records of some of these species have not been checked; and d) the absence of recent records can be the result of great rarity.

For all of these reasons it is very difficult to indicate how many species have become extinct. This is illustrated by the fact that of the 37 species which Barkman (1958) concluded had disappeared, 6 were found during the WHEN survey (i.e., *Calcium abietinum*, *C. salicinum*, *Chaenotheca trichialis*, *Enterographa crassa*, *Parmelia quercina*, and *Pyrenula nitida*). Nineteen species can be said with a rather high degree of certainty to have disappeared, but there are probably many more as well. This question and species which are new for The Netherlands will, however, be reported elsewhere by authors who participated in the WHEN project.

15. COMPARISON WITH THE SITUATION AROUND 1950

It is difficult to describe changes since 1950 precisely. Barkman's study was concerned with the ecology and phytosociology of epiphytes rather than the determination of epiphytic richness and geographical differences in it. As a result, his observations are mainly restricted to th vegetations of special interest for his investigation, and do not provide species lists made as complete as possible for given regions. Furthermore, his records include less common species, but floristic data on common species are not complete.

The maps of the 'epiphyte deserts' in The Netherlands published by Barkman in 1958 and 1969 are based on the impressions he obtained during his field work, and on his distribution maps for both epiphytic lichens and epiphytic mosses not only for free-standing trees but also from parks and forests. A region was considered an epiphyte-desert by him if the local epiphyte vegetation was appreciably poorer than would be expected from the findings in comparable areas. Consequently, the various epiphyte deserts he shows are not equally poor. In 1969, Barkman made an attempt t correct this to some extent.

It is difficult to compare the situation around 1950 with the presen one on the basis of the number of species per square, because the picture given by the 1950 data is too incomplete. The distribution of the species serves this purpose better: it can be stated with a reasonable degree of reliability whether a species still occurs in squares in which Barkman found it. The chance that 'disappearance' depends on differences in observation and on chance is smaller when the species are taken together in groups.

For this purpose, a map was prepared showing the sensitivity classes of the species for all the squares studied by Barkman for which suitable data were available (Appendix 106). This map is based on Barkman's original notes.

Comparison of this data with the present distribution (see also Appendix 79) shows where sensitivity classes have disappeared (e.g. where species of class H still occurred in 1950, the most sensitive species now present belong to class G). Here, _Anaptychia ciliaris_, _Physc' aipolia_, and _Parmelia perlata_ are considered to belong to sensitivity class H (see Chapter 14); although no sensitivity class was determined for these species, placement in class H is justifiable on the basis of the highly ndicative data concerning the earlier and the present distribution.

On this basis only decline can be determined, because the chance is elatively small that the absence of a sensitivity class in the WHEN surey is a matter of chance or incomplete observation, whereas this chance s much greater for the pre-1950 data.

Appendix 107 shows the squares in which reduction has occurred. The ollowing conclusions can be drawn on this point (see also App. 12, 79, nd 106):

) the currently large epiphyte-poor area in the western part of The Netherlands has become appreaciably poorer since 1950 mainly along the periphery. The sensitive species formerly still found south of Amsterdam have disappeared, which indicates that although epiphyte-rich places still occurred between Amsterdam and Rotterdam/'s-Gravenhage around 1950, this area now forms a single epiphyte-poor region.

) Whereas in 1950 sensitive species were only absent in the immediate vicinity of the city of Nijmegen, this area has now become much larger; to the west of Nijmegen and around and northeast of Arnhem, all of the sensitive species have disappeared and the area is now very poor in epiphytes.

) The epiphyte vegetation of Zeeuws-Vlaanderen, which was already poor in 1950, has become still more impoverished. The island of Zuid-Beveland shows a sharp decline, and the sensitive species have also disappeared from the western part of Brabant.

) In Noord-Holland (west of Hoorn) and in the region between Amsterdam and Utrecht, which are still among the richest areas in The Netherlands as far as the number of species is concerned, the sensitive species (class H) have disappeared (see also Chapter 12).

) In Friesland, here and there in Groningen, and on the coast of Zeeland, there has also been a reduction, although these provinces still have squares in which sensitive species occur.

The northern part of the province of Groningen does not seem to have ndergone impoverishment, but according to Barkman (personal communication) he epiphytic mosses have declined appreciably.

The map in Appendix 107 can be roughly divided into three regions with ifferent degrees of decline:

Region A, comprising the Wadden islands (except Texel) and the north-eastern part of the mainland: limited decline; ratio between numbers of squares with ($\overset{\bullet}{\circ}$) and without (x) decline 7/48 = 0.15

Region B, coastal zone and a strip along the west and south border of Region A: squares with and without decline in a ratio of 40/27 = 1.48.

Region C, the central and southern parts of the country: decline in most of the squares; ratio 45/7 = 6.43.

In addition to this comparison, an attempt was also made to prepare a map of the 'epiphytic deserts' around 1950; the results (App. 108) can, with some reserve, be compared with the map of the present situation (App. 30).

This map was made in the following way:

I. To the squares for which suitable data were available for the period around 1950 a class was assigned (on the basis of classification T_4). Comparison with the map of the epiphyte deserts (App. 1) leads to the conclusion that the deserts correspond roughly with classes 1-3. But some of these areas were designated as a 'desert' at a higher degree of epiphyte richness. This concerns particularly areas for which few data were available and which lay between two distinctly epiphyte-poor places. This holds, for instance, for the areas north of Amsterdam and the eastern part of the province of Utrecht.

II. Furthermore, the present study has shown that in areas belonging to classes 1-3, H and G species occur very rarely and F species only sporadically.

On the basis of these two considerations, the old epiphyte-desert map (App. 1) and the map showing the occurrence of the sensitivity classes (App. 106) were used to make a new map of the situation around 1950 (App. 108). The borderlines on this map are uncertain, however, due to the above-mentioned differences in observation and because no data at all are available for many squares. If desired, this map can be compared with the one in App. 30; such comparison shows in particular that the epiphyte-poor areas in Zuid-Holland and around Amsterdam have expended until they merged, that the areas around Arnhem and Nijmegen have become larger, and that Zeeland and western Brabant have become poorer. But comparison on this basis is much less reliable than the much more direct comparison on the basis of the disappearance of species.

Conclusions: The epiphyte-poor areas where only a few epiphyte lichens occur have become appreciably larger since 1950. The sensitive species have disappeared from a large number of places. Only for Drenthe, northern Overijssel, the eastern part of Friesland, and the Wadden Islands (except Texel) has no decline of the epiphyte vegetation been demonstrated.

16. FUMIGATION EXPERIMENTS

16.1 Introduction

In addition to the survey described in the preceding chapters, a number of fumigation experiments were performed. The experiments on the effect of air pollution on lichens reported in the literature can be divided into three groups:

a. Transplantation experiments: Healthy thalli were transferred to places where air pollution was known or expected to exist, to find out whether this change would lead to deleterious effects such as discoloration of the thallus surface or degradation of chlorophyll (see e.g. Brodo 1961; Rao and LeBlanc 1965; LeBlanc and Rao 1966, 1973b; Schönbeck 1968, 1972; LeBlanc, Comeau and Rao 1971; Nash 1971; Horntvedt 1975; Jürging 1975).

b. Fumigation experiments in glasshouses: Thalli were exposed to SO_2 in aqueous or gaseous form, and the effect on the respiration, photosynthesis, and the chlorophyll content was investigated (see e.g. Pearson and Skeye 1965; Hill 1974; Nash 1971, 1973; Baddeley, Ferry and Finegan 1973; Richardson and Puckett 1973; Puckett, Richardson, Flora and Nieboer 1974; Türk, Wirth and Lange 1974; Wirth and Türk 1975; Türk and Wirth 1975).

c. Fumigation experiments in the field: Thalli were exposed in the field to an artificial source of a gaseous pollutant (see Skye 1968; Jürging 1975).

The disadvantage of transplantation experiments is that the conditions are often not sufficiently known. The fumigation experiments can be objected to on the grounds that so far only high concentrations applied for short intervals have been used, so that it is not known whether the results hold for more realistic low concentrations over a longer period. Furthermore, most of the fumigation experiments in glasshouses have almost invariably been done with sulphur dioxide or sulphite. All this explains why we welcomed the chance to include lichens in fumigation experiments performed by the Instituut voor Plantenziektekundig Onderzoek (IPO) at Wageningen. These experiments concerned prolonged fumigation with various cases at low concentrations.

16.2 Methods

In these experiments the concentration of the gases was very low, comparable with those prevailing in polluted areas, and the duration of the exposure was very long (see Appendices 109 and 110). For this and technical reasons, the method applied for field transplantations was used. This method was developed by Schönbeck (1968). Samples of the species of lichens to be investigated were removed together with the underlying bark from the trunk or branch of the tree with a circular knife (diameter 4.5 cm), and attached to small boards with a grafting wax. The lichen species were placed randomly in the holes punched in the boards, but the distribution was the same for each kind of exposure. Before, once or twice during, and after fumigation, all samples were photographed (35 mm diapositive film; Kodachrome II in the first series, otherwise Agfachrome 50 S). The reproduction ratio was 1:2 throughout. Illumination was given by a circular electronic flash.

As the measure of any deleterious effect of fumigation, we used the percentage of the thallus surface with a distinct change in colour as compared with the pre-fumigation colour. This was determined by simultaneous projection, with two identical projectors, of the pictures taken before and after the exposure to a given gas, from which the percentage damaged (discoloured) area was estimated.

The fumigations were performed in the glasshouses of the IPO at Wageningen (see Appendix 114). Because the experiments in which our material could be included concerned higher plants (bulb, tree, and bush species) and due to technical limitations, the microclimatological conditions were not ideal for lichens in that the relative humidity was on the low side.

Four series of fumigations with a duration of 3 to 6 months were performed. These gases were HF (1st and 2nd experiments); SO_2, O_3, and C_2H_4 (3rd experiment); and SO_2, O_3, and $SO_2 + O_3$ (4th experiment). A control group of lichens was also placed in an untreated glasshouse.

Data on the concentrations and other conditions are shown in Appendices 109 and 110. Hydrogen fluoride (HF) was applied for about 80% of the 24-hour period. The HF concentrations were determined with impingers; later, other methods showed that the greater part of the HF in the untreated glasshouse derived from fluor in dust particles. Therefore, the latter may also have been present in the glasshouse used for fumigation.

During the treatments in the first and fourth experiments the boards were only sprinkled when the automatic installation was turned on for the other plants (once or twice a week). In the second experiment half of the samples were sprinkled one or two extra times a week (the 'wet' samples), and in the third experiment this occurred for all of the samples.

To evaluate the effect of the transplantation itself, during the first and second fumigation experiment a group of samples were taken and set out in the same place (in the Speulderbos, indicated in Appendices 111 and 112 as forest). *Evernia* was also placed there during the first experiment, and during the second this species was transplanted to a place with a closer resemblance to its original site, i.e., open ground near the RIN in Leersum (indicated as field). In the second experiment the other species were also transplanted to both the forest and field sites.

Immediately after the fumigation of the first experiment, the samples were transferred from the glasshouse to the Speulderbos to find out whether recovery can occur. For the samples exposed to high concentrations and half of the control samples (i.e., from one of the two untreated glasshouses), this transfer was made after 75 days of fumigation and for the samples exposed to a low HF concentration and those from the other untreated glasshouse after 137 days.

In the first experiment the lichens were photographed before the experiment and 75, 137, and 241 days after the start; in the second before the start and after 56 and 84 days, in the third before and 170 days after the start; and in the fourth before the start and after 180 days.

The material for the fumigation experiments was collected and processed at most two days before the start of the experiment.

The number of samples in each fumigation is shown in Appendices 111, 112, and 113. The lichen samples were collected in the indicated places (see also Appendix 114):

Buellia canescens (4th experiment): from willows along a road near Leersum

Cetraria glauca (1st, 2nd experiment): oaks in the Speulderbos (municipality of Ermelo)

Evernia prunastri (1st, 2nd, 3rd experiment): oaks along a road between Leersum and Scherpenzeel; (4th experiment): oaks along a road near Maarsbergen

Parmelia saxatilis (all 4 experiments): oaks in the Speulderbos

Parmelia sulcata (2nd experiment): Speulderbos; (3rd experiment): apple trees in Glimmen; (4th experiment): elms along a road near Leerdam

Parmelia physodes (1st, 2nd experiment): oaks in Speulderbos; (3rd experiment): oaks in Speulderbos and apple trees in Glimmen; (4th experiment): oak thicket near Leersum

Physcia tenella (3rd experiment): apple trees in Glimmen; (4th experiment): willows along a road in Langbroek

Xanthoria candelaria (4th experiment): willows along a road near Leersum

16.3 Results

The results are given in Appendices 111-113. The significance of differences in damage to the fumigated samples was estimated by comparison of damage data for fumigated samples with those for the controls on the basis of the Mann-Whitney test (Sokal and Rohlf 1969), taking the limit of significance at P=0.2.

16.3.1 Hydrogen fluoride (see Appendices 111 and 112)

Cetraria glauca: The high concentration had a limited but significantly injurious effect in the first experiment. The low and very low concentrations did no damage.

Evernia prunastri: No effect was seen except under prolonged exposures (137 days). Under shorter exposure, only the high concentration applied to the wet samples gave some damage. Transplantation to the forest, a very different habitat from that of the sampling site was not supported by this species. It was noteworthy that the samples exposed to the high concentrations for 76 days were not in a worse condition immediately after the fumigation than the controls but died off significantly more rapidly than the controls after transplantation to the forest.

Parmelia physodes: The high concentration caused significant damage in the first experiment but not in the second (where the controls showed severe damage). After exposure (1st experiment), the fumigated material transferred to the forest died, but the controls showed little mortality. The low concentration (1st experiment) and the very low concentration (2nd experiment) had no (significant) effect. Controls transplanted to the forest after 137 days died, but those held in the untreated glasshouse for only 75 days did not.

Parmelia saxatilis: In the first experiment and the wet samples of the second, the high concentration caused damage. The low concentration (1st experiment) and the very low concentration (2nd experiment) did not lead to significant damage. The dry samples in the second experiment showed slightly less damage than the controls, but the difference was not significant. The controls died after transplantation to the forest, whereas the samples placed in the forest at the beginning of the experiment showed much less damage.

Parmelia sulcata: The number of samples of this species was too small to permit the drawing of conclusions.

Conclusions

It is clear that at the high concentration (2 ppb) in the first experiment HF had a negative influence on all of the species, although for Evernia this only became manifest after the fumigation. The effect was less distinct in the second experiment. The controls too showed a very high percentage of damage. It seems possible that for Evernia the duration of exposure in the second experiment was too short to cause visible damage In all of the species except Evernia the low concentration (0.46 ppb) cuased damage in the first 75 days, although the divergence from the controls was not significant. But after 137 days the damage had become smaller or the same (but not significantly) as the damage in the controls. Only Evernia was significantly damaged by this concentration, and only after 137 days. The very low concentration gave no demonstrable effect.

No explanation can be offered for the finding that the damage in the second experiment was much higher in the controls than in the first experiment (see also section 14.4). If the damage to the controls occurred because the lichens were too briefly capable of physiological activity (e.g. due to an insufficient amount of or too short a period of exposure to moisture), this could also have reduced the effect of the fumigation as compared with the effect under more favourable conditions, because the effective duration of the fumigation was shorter.

The damage to the fumigated samples especially those exposed to the high concentration, was almost always greater in the wet than in the dry specimens.

16.3.2 Ethylene (see Appendix 113)

In Evernia prunastri, Parmelia saxatilis, and Physcia tenella, C_2H_4 caused very distinct and highly significant damage. In Parmelia physodes the damage was also higher than that in the controls. Only in P. sulcata did the damage differ hardly at all from that in the controls.

16.3.3 Ozone (see Appendix 113)

In the fourth experiment, O_3 had a significant damage-limiting effect on two species, Buellia canescens and Parmelia physodes. In the third experiment this effect was not visible in P. physodes, but there was hardly any control damage either and the O_3 concentrations were slightly lower. In two species (Evernia prunastri and Physcia tenella) O_3 had a significantly negative effect, but in Ph. tenella this occurred only in the fourth experiment (with slightly higher concentrations and, in view of the control mortality, less favourable conditions for this species). In Parmelia sulcata O_3 fumigation gave more damage, but not significantly so. No visible effect was seen in P. saxatilis or Xanthoria.

16.3.4 Sulphur dioxide (see Appendix 113)

The effect of SO_2 was damaging for Evernia prunastri and Parmelia sulcata in both experiments. In the third experiment P. physodes, P. saxatilis, and Physcia tenella showed significantly more SO_2 damage compared with the controls, but in the fourth it was not significantly larger than the considerable control damage. For Buellia and Xanthoria the SO_2 damage was not greater than for the controls, which were severely affected too.

16.3.5 Sulphur dioxide combined with ozone (see Appendix 113)

This combination gave more damage than was seen in the controls (in three cases significantly more). Only Parmelia saxatilis showed no divergence. For Buellia canescens, in which O_3 had a damage-limiting effect, the combination gave more damage than SO_2 alone and more than occurred in the controls. For P. physodes, in which O_3 is also damage-limiting, the damage caused by the combination lay between that caused by the gases separately. In Physcia tenella the damage was also intermediate. For Evernia prunastri, Xanthoria candelaria, and P. sulcata, the combination gave slightly more damage than either component alone; in P. saxatilis the effect was about the same.

16.4 Discussion

Comparison of the present results with those published in the literature shows some striking points. A sensitivity rank order for the species cannot be derived from our findings, in part because of the wide variation in the reaction to the various fumigations. It is also relevant here that the damage to the controls was very high in some cases and was not the same for all species. It is also clear that the sensitivity of a given species varies with the prevailing conditions (see remarks on Physcia tenella and Evernia prunastri in section 13.3 and also Baddeley et al. 1973). The often substantial control damage, which also varied widely between species, shows that the glasshouse conditions were far from ideal. This could also have increased the intra-species variation of the effect of the fumigation. The occurrence or absence of such variation is seldom mentioned in the literature. Barkman et al. (1969) state that in Gilbert's work sample variability did not exceed 10%.

Besides the differences between specimens of one species collected in one place there are also great differences between samples of one species collected in different places, as can be seen for P. physodes in the third experiment. The latter differences are perhaps attributable to differences between populations with respect to optimal conditions (e.g. as to the light compensation point; Pearson 1973).

The appreciable difference in control damage between experiments is also striking; in the second experiment this damage was higher than in the other experiments. This difference might be attributable to the fact that the second experiment ran from January to April, whereas the others started in May, as a result of which the difference in climate between normal conditions in the field and those in the glasshouse were much greater in the second experiment than in the others. The limited control damage during the third experiment was probably due to the additional sprinkling of the lichens.

It is clear from the literature that lichens are more sensitive under moist than under dry conditions (Türk et al. 1974; Nash 1971), and this is confirmed by the second experiment in which the damage due to the fumigation was greater for the 'wet' samples than for the 'dry' ones.

Evernia prunastri tolerated the conditions in the glasshouse better than the other species. This species was indeed sampled in a much drier habitat than the others. On transfer to the moist dark conditions of a forest, Evernia died, and the wet control samples of this species showed more damage than the dry ones. This ability to endure the glasshouse climate was accompanied by a distinct reaction to the fumigations, which makes the species highly suitable for this type of experiment.

Although, for the above-mentioned reasons, the results for the various species can hardly be compared, it is an interesting finding that under conditions which were more or less optimal for a species, the low concentrations of the gases had such a strong effect. The fact that this low concentration -which occurs in the field- causes severe morphological changes makes it probable that the low epiphyte richness found in the survey could have been caused by air pollution. At the low concentrations used in the glasshouse higher plants were also affected, as the IPO data show (Mooi 1974, Spierings 1974). The literature on the effects of air pollution on high plants usually concerns higher concentrations applied for short periods (see e.g. Air Pollution 1969); prolonged fumigation has seldom been applied (see e.g. Feder 1970).

The effective (average) concentration and duration of exposure for the lichens in the present experiments are difficult to estimate. Comparison with situation in the field is also difficult, because a) the frequency distribution of the concentration is very different in the glasshouse, and b) the time during which the lichen is physiologically active will be very different because of the divergent climatological conditions. This too influenced the comparability of the results of the various experiments.

The HF concentrations applied were very low; even the 'high' concentration of 2 ppb (for 80% of the 24-hour period) occurs regularly in the field. In previous fumigation experiments (Nash 1971; Comeau and LeBlanc 1972) much higher concentrations were applied.

The damage caused by SO_2 is very clear; the concentrations used were also rather high. The 'day averages' of these fumigations were 50 to 60 ppb, whereas according to the survey results, epiphyte zone A occurs when 50% of the days have more than about 35 ppb (100 $\mu g/m^3$) as day average. This concentration is, however, small compared with those reported in the literature for fumigation studies.

Nothing was known previously about the effect of O_3 and C_2H_4 on lichens. The strong effect of ethylene observed in the present study is striking, although the concentration was rather high. In the field such concentrations probably only occur locally and for very short periods. The effect observed for ozone is very interesting, especially the damage-limiting effect seen in two species. A study done by Comeau and LeBlanc (1971) has shown that ozone has a stimulatory effect at low concentrations and an inhibitory effect at high concentrations (on *Funaria hygrometrica*). The concentrations of ozone applied in our study occur commonly in the field. On the basis of our findings, further investigation of the effect of ozone alone and in combination with other gases such as sulphur dioxide seems worthwhile.

Taken as a whole, the present experiments show clearly that fumigation experiments can supplement findings in the field, but that sufficient attention must be paid to obtaining optimal conditions comparable with those of the natural habitat of the lichens. The wide divergence of these conditions among the various species increases the technical difficulties involved in such experiments and the problems encountered in interpreting the results. Considerable attention must also be paid to fluctuations in these conditions, not only with respect to climatological factors with an influence on, for instance, water content (Farrar 1973, Richardson and Puckett 1973), but also with respect to the changes in the concentration of the gas (Puckett *et al*. 1974).

17. DISCUSSION OF THE USE OF LICHENS AS INDICATORS OF AIR POLLUTION

17.1 Correlation with SO_2 levels -- comparison with the results of other authors

The results of the WHEN project have been discussed in detail above. Comparison of the correlation between SO_2 levels and epiphyte richness found in his study and that found by other authors involves difficulties. None of the other authors used the median concentration (50 percentile); most of them employed the average concentration for either the winter period or the entire year.

To facilitate comparison somewhat, the correlation was determined between the average (winter or year) concentrations and the 50 percentile used in the pesent study. According to the results: the median winter concentration = average concentration -10 (in $\mu g/m^3$). This was arrived on the basis of the data for two winters recorded in the area of the Rhine estuary (r=0.96).

For the relationship between the winter and the year values, it holds for the 50-percentile that:

In Noord-Holland:year value = 0.70 x winter value -8 $\mu g/m^3$ SO_2

In Zuid-Holland :year value = 0.70 x winter value +5 $\mu g/m^3$ SO_2

This implies that there are greater differences between the winter and year concentration in Noord-Holland than in Zuid-Holland, which could be the result of a higher proportion of less season-dependent or non-season dependent sources of SO_2 in Zuid-Holland.

Comparison is also difficult because the present study has shown that in The Netherlands the peak values (98 percentile) are important with respect to the epiphyte vegetation. It is noteworthy than a relationship between peak SO_2 values and the richness of the epiphyte vegetation is not mentioned anywhere in the literature. This is a new result, not previously reported. Hawksworth (1973) even says that: '... it is not surprising that no correlations with maxima have so far been demonstrated'. Now, maxima are not the same as 98 percentiles, because the latter indicates above which average day-concentration 2% of the days in the period in question are lying.

The difference between Zuid-Holland and Noord-Holland in itself indicates how very difficult it is to compare winter medians with average (year) concentrations found in some other places.

Comparison is only possible for zones A and B, because the 50 percentile is less important in richer zones. Low 50 percentiles (*ca.* 30-70 $\mu g/m^3$) can occur (as found in The Netherlands) in combination with highly divergent 98 percentiles, and in these cases it is the latter which determine how many epiphytic lichens can still occur (see also App. 41,43, 71, and 73).

Taken generally, the zones in the various countries corresponds as follows:

SO_2 concentrations given as	Zones with corresponding SO_2 concentrations (in $\mu g/m^3$)		
The Netherlands	Zone A	Zone B	Zone C
50 percentile winter	>100	*ca.* 70-100	*ca.* 50-70
Great Britain (Hawksworth & Rose 1970)	*ca.* zone 1+2	*ca.* zone 3	*ca.* zone 4-5
mean winter concentration	150	125	60-70
corresponding 50 percentile winter	140	115	50-60
Germany (Jürging 1975)	Krustenzone		Anflugzone
"grob gesehen durchschnittlich"	100-150		50-100
Ireland (Fenton 1964)			
(year?) average		0.018 ppm	
corresponding winter median		*ca.* 60	

On this basis, the SO_2 level at which the first foliose species appear (zone C) seems to be about the same in all four countries. The concentrations mentioned by Hawksworth and Rose for the poorer zones (particularly zone 3) are on the high side compared with the concentrations in The Netherlands.

17.2 Other air pollutants

It is known from both field and laboratory findings that various substances which may pollute the air have a deleterious effect on lichens. For HF, reference is made to Gilbert (1973). The fumigation experiments described in Chapter 16 showed that appreciable damage can occur even at low HF levels. The same was found for ethylene and ozone. The influence of heavy-metal pollution, radionuclides, radiation, smoke, and fertilizers is discussed by James (1973). It may be concluded that when an impoverished epiphyte vegetation is found, its condition can be caused by other components of polluted air besides sulphur dioxide, and that whithout any further information sulphur dioxide cannot be held solely responsible.

17.3 Other conditions

As already mentioned in the relevant chapters, the sensitivity of a lichen species varies with such conditions as the species of phorophyte on which it occurs, and there are conditions (e.g. moisture) under which the effect of air pollution is enhanced or diminished. Thus, to permit comparison these conditions much be as much the same as possible. This can be attained by standardizing the sampling stations as much as possible. Furthermore, conclusions concerning the sensitivity of species and correlations with air pollution drawn from research done in other countries cannot be applied without further analysis.

In addition, it is necessary to take into account such factors as the absence of suitable phorophytes and the age of the area (or of the trees in an area), as, for instance, in a newly reclaimed area like the Noordoost Polder.

17.4 The previous history of an area

It must be kept in mind that the observed distribution of epiphytic lichens is the result of conditions and events over a long series of years. In this respect, two aspects of the problem are of special importance.

17.4.1 Trends in the sulphur dioxide levels

These trends can take several forms:

a. The levels of SO_2 can remain roughly the same over a period of many years. In that case the epiphyte vegetation has probably reached an equilibrium with the prevailing concentrations. Under such conditions the correlation found between SO_2 and the epiphyte vegetation approaches the real relationship closely.

b. The SO_2 levels can show a rising trend. If the reaction of the epiphyte vegetation is slow and equilibrium with a given pollution situation is only reached after some years, the epiphyte level will lag behind the pollution level.

c. The SO_2 levels van show a tendency to decrease. When the epiphytes react slowly, the vegetation will be poorer than would be expected in an equilibrated situation.

Unfortunately, information about a trend is known for the SO_2 concentrations at only a few of the stations. In the Zuid-Holland region, it is known for a few stations in Rotterdam that the levels dropped by 10-40% between 1965 and 1970 in connection with the change-over to natural gas. These were all stations with high to very high concentrations. For the Westland region, where vegetables are raised in glasshouses, the fuel change-over began after 1971 and both the lichen and the SO_2 findings date from just before and the early period after the change. During the roughly five years before the survey, the rather low concentrations recorded on the island of Voorne do not show a distinct trend. In the same period there was a slight increase in Noord-Holland except in Amsterdam, where the high concentrations began to decline. The Sloe area showed a slight decrease.

The difference in trend from place to place is probably responsible for the variation in the relationship between SO_2 levels and epiphyte richness. Thus, the true epiphyte richness will sometimes differ from the predicted. The degree of this divergence is also dependent on the rapidity with which the lichens react.

17.4.2 Reaction rate of lichens

There is relatively little information available on this point for either disappearance or reappearance. The very close correlation between the absence of Lecanora expallens and SO_2 levels suggests that this species reacts rapidly to both increase and decreases in the SO_2 concentration. The rate at which disappearance occur will also be directly related to the degree of the rise in SO_2 level, for instance, of peak values in relation to the average concentrations. It is also possible that the reaction is dependent on conditions prevailing in the preceding period: a thallus already affected by rather high concentrations but not yet fatally will perhaps be killed by an SO_2 concentration that would not be fatal for a healthy thallus. Furthermore, as the fumigation experiments have shown, there is also variation in the reaction rate within a species.

With respect to the recolonization after reduction of air pollution, it is possible that such a return occurs at quite different pollutant levels than those leading to disappearance. Margot (1973) has shown that the germination capacity of the soredia decreases under the influence of SO_2. In this respect these factors are of importance: a) the dispersion and juvenile stages of a lichen (spores, soredia, isidia, and young thalli) do not necessarily have the same sensitivity to pollutants as older thalli. Research results obtained in many species of plants belonging to other groups make it seem likely that the sensitivity of these stages is greater than that of adult thalli. b) Accessibility also plays a role here. When a species disappears from a large area under the influence of air pollution, it will take longer for that species to return to the whole area after the pollution decreases than is the case when it has disappeared from only a small area. In this connection it is also important that when air pollution is limited a species often does not disappear but becomes sterile. This too will limit the increase of a species under decreasing air pollution (or promote a decrease under constant air pollution), especially when the species cannot disperse vegetatively, and c) complete recovery of the epiphyte vegetation can take a very long time, the factors forming obstacles including irreversible changes in tree bark and the (virtually) complete extinction of species.

The only report of a return of lichens after a drop in SO_2 levels concerns München (Jürging 1975). After 1969 there was an appreciable improvement in the condition of the air there, and in 1973 small thalli were seen again, mainly for _Parmelia physodes_ but also for _Parmelia sulcata_, _Xanthoria parietina_, _Physcia_ spp., and very infrequently for _Usnea_ spp., Parmelia _furfuracea_ (twice), and _P. caperata_ (once). This shows that a few species can recur within a few years.

It can be clearly seen from the situation in the Noordoost Polder, for instance, that the formation of a rich epiphyte vegetation requires much time. In this new polder, where the first trees were planted around 1944, the level of air pollution is low. Nevertheless, various species are entirely absent there, for example _Parmelia acetabulum_, _P. subrudecta_, and _Buellia canescens_. The number of species in the west of this polder (along the shore of the IJsselmeer) is also still low (see section 12.5).

It is not *a priori* expectable that the sequence of return will be the exact reverse of the sequence of disappearance, firstly because the accessibility is not the same for all lost species, and secondly because the juvenile/adult sensitivity ratio is not the same for all species.

17.5 Conclusions

The results of this study show that:

1. At very high SO_2 levels the distribution of a few sensitive species shows good correlation with these concentrations.
2. At lower SO_2 levels the correlation of the distribution of the individual species is less distinct and is best for the number of species per sampling site.
3. In still less polluted areas the best correlation is obtained for the number of species per square.
4. Where the SO_2 levels are extremely low, the distribution of very sensitive species is also of importance.

These points must be taken into consideration when conclusions about SO_2 concentrations are based on epiphytes.

It is also of importance that epiphyte impoverishment can be caused by many other factors as well as SO_2. This is especially the case in areas with lower SO_2 concentrations. Here, the peak SO_2 values are particularly important. Our results do not support Hawksworth's (1973) statement that 'A rapid and accurate estimate of mean annual or mean winter sulphur-dioxide levels [can be] based solely on the lichen vegetation'.

If, for instance, a labour-saving repeat survey is restricted to a few species, the best procedure would be to select the most common species (one or more) from each sensitivity class, because the occurrence of these species is influenced the least by interfering factors (see App. 84).

17.6 Comparison with the norms for sulphur dioxide immission

In The Netherlands recommended norms have been formulated by the Gezondheidsraad (1971). Limiting values were given for relatively clean areas (norm A), for industrialized and urbanized areas (norm B), and for areas with currently distinct air pollution (norm C). The limiting values for these norms are expressed as the 50 and 98 percentile of the frequency distribution of the 24-hour values of a whole year. For the most severe

norm, i.e. norm A, the 50 percentile value is 75 $\mu g/m^3$ and the 98 percentile value 250 $\mu g/m^3$ SO_2.
(The A, B, and C norms are not related in any way to epiphyte zones A, B, and C. Norm A is strictest, i.e., prescribing the lowest concentrations of SO_2; zone A is the poorest in epiphytes.)

Comparison with the limiting values found for the occurrence of epiphytes involves the problem that these values pertain to the frequency distribution for the six winter months. The relationship between annual and winter values for the 50 percentile has been discussed in section 17.1; for the 98 percentile, the winter value amounts to about 1.1 times the annual value. In view of the spread, an annual median value of 75 $\mu g/m^3$ corresponds with a winter median value of 80-120 $\mu g/m^3$, and for the 98 percentile an annual value of 250 $\mu g/m^3$ corresponds with a winter value of 260-320 $\mu g/m^3$.

It can be seen from Appendix 73 that at these concentrations an epiphyte vegetation belonging to zone A (class 1) or zone B (class 2) is to be expected. This means that a SO_2 level considered acceptable according to the strictest of the recommended norms, virtually no epiphytic lichens can survive.

SUMMARY

The WHEN workgroup inventoried the epiphytic lichens on free-standing trees troughout The Netherlands, using a 5 km square-grid as a basis. The 1,395 squares were divided into six classes of epiphyte richness. This was done on the basis of the number of lichen species per tree species per square. The results are given in a map of epiphyte richness in The Netherlands (Appendix 12).

The data for the west of the country were analysed in more detail, resulting in a zone map (Appendices 57 and 64).
It proved possible to construct a good descriptive model of the correlation between SO_2 concentrations and epiphyte richness (Appendix 73).

a. At very high SO_2 concentrations only a few relatively insensitive lichen species occur. This area coincides with median winter SO_2 concentrations higher than 100 $\mu g/m^3$.
b. At lower concentrations, winter peak values (98 percentile: the concentration below which 98% of the winter day values lay) are more important.
c. At high concentrations, SO_2 is the limiting factor, at lower concentrations other factors are important too.

Lichen species found in 25 or more squares have been ranked according to sensitivity. As a measure of sensitivity use was made of the extent to which a species is restricted to epiphyte-rich squares. The resulting rank order (Appendix 77) is independent of the classification of the squares.

The distribution maps of 20 lichen species are given. Especially the more sensitive species proved to have disappeared from large areas of The Netherlands. Comparison of the present situation with data from around 1950 reveals a decline almost everywhere in the country. Epiphyte-poor areas have grown larger, and sensitive species disappeared from less poor areas. The southern and western parts of the country are impoverished, the area between Rotterdam and Amsterdam and the area between Arnhem and Nijmegen in particular have become very poor (Appendix 107).

Four fumigation experiments were performed, using HF, SO_2, C_2H_4, O_3 and SO_2 in combination with O_3 at realistic concentrations and prolonged exposition (75-180 days). It was found that at the low concentrations applied and under glasshouse conditions, these air pollutants cause significant morphological damage to all or some of the tested lichen species.

REFERENCES

AIR POLLUTION (1969) Proceedings of the First European Congress on the Influence of Air Pollution on Plants and Animals, Wageningen 1968. Wageningen, Pudoc

ATLAS VAN NEDERLAND; Samengesteld door de Stichting Wetenschappelijke Atlas van Nederland (1963-) 's-Gravenhage Staatsdrukkerij en Uitgeverij.

BADDELEY, M.S., FERRY, B.W. & FINEGAN, E.J. (1973) Sulphur dioxide and respiration in lichens. In: Ferry, B.W., Baddeley, M.S. & Hawksworth, D.L. (eds.) Air Pollution and Lichens 299-313. London, The Athlone Press.

BARKMAN, J.J. (1958) Phytosociology and ecology of cryptogamic epiphytes. Assen, The Netherlands, Van Gorcum.

BARKMAN, J.J. (1963) De epiphyten-flora en -vegetatie van Midden Limburg Verh.K.ned.Acad.Wet. II,54(4),1-46

BARKMAN, J.J. (1969) The influence of air pollution on bryophytes and lichens. In: Air Pollution, Proceedings of the First European Congres on the Influence of Air Pollution on Plants and Animals. Wageningen 1968, 197-209. Wageningen, Pudoc.

BARKMAN, J.J., DOING, H. & SEGAL, S. (1964) Kritische Bemerkungen und Vorschläge zur quantitativen Vegetations analyse. Acta Botanica Neerlandica 13, 394-419.

BARKMAN, J.J., ROSE, F. AND WESTHOFF, V., reporters (1969) Discussion in Section 5: The effects of air pollution on non vascular plants in: Air Pollution, Proceedings of the First European Congress on the Influence of Air Pollution on Plants and Animals, Wageningen 1968. 237-41.

BRAAK, C. (1950) Het klimaat van Nederland. Den Haag: Servire

BRAND, A.M. (1976) Epifytische lichenen en luchtverontreiniging in Zuid-Holland. (not published).

BRODO, I.M. (1961) Transplant experiments with corticolous lichens using a new technique. Ecology, 42,838-41.

COMEAU, G. & LEBLANC, F. (1971) Influence de l'ozone et de l'anhydride sulfureux sur la régéneration des feuilles de Funaria hygrometrica Hedw. Naturaliste can.,98, 347-58.

COMEAU, G. & LEBLANC, F. (1972) Influence de fluor sur le Funaria hygrometrica et l'Hypogymnia physodes. Can. J. Bot. 50,847-56.

COPPINS, B.J. (1973) The 'Drought Hypothesis'. In: Ferry, B.W., B.W., Baddeley, M.S. & Hawksworth, D.L. (eds.). Air Pollution and Lichens. 124-42. London, The Athlone Press.

DE SLOOVER, J. (1964) Végéteaux épiphytes et pollution de l'air. Revue Quest.scient. 25, 531-61.

DE SLOOVER, J. AND LEBLANC, F. (1968) Mapping of atmospheric pollution on the basis of lichen sensitivity.
In: Misra, R. and Gopal, B. (eds.) Proc.Symp. Recent Adv. Tropical Ecology. 42-56.
Varanasi, The International Society for Tropical Ecology.

DE SLOOVER, J. ET LEBLANC, F. (1970) Pollutions atmosphériques et fertilité chez les mousses et chez les lichens epiphytiques. Bull. Acad. Soc. Lorr. Sci. IX, 82-90.

DE WIT, TOKE (1976) Flechtenverbreitung in den Niederlanden und ihre Abhängigkeit von der Luftverunreinigung. Schriftenreihe für Vegetationskunde, 10 (in press).

DOMRÖS, M. (1966) Luftverunreinigung und Stadtklima im Rheinisch-Westfälischen Industriegebiet und ihre Auswirkung auf den Flechtenbewuchs der Bäume.Arb.Rhein.Landesk. 23,1-132.

FARRAR, J.F. (1973) Lichen physiology progress and pitfalls In: Ferry, B.W., Baddeley, M.S. & Hawksworth, D.L. (eds.) Air Pollution and Lichens. 238-82, London, The Athlone Press.

FEDER, W.A. (1970) Plant respons to chronic exposure of low levels of oxidant type air pollution. Environ. Pollut. 1,73-9.

FENTON, A.F. (1964) Atmospheric pollution of Belfast and its relationship to the lichen flora. Ir. Nat. J. 14,237-45.

FERRY, B.W., BADDELEY, M.S. & HAWKSWORTH, D.L.(eds.) (1973) Air Pollution and Lichens, London, The Athlone Press.

GEZONDHEIDSRAAD (1971) Advies inzake grenswaarden SO_2, Verslagen en Mededelingen, Ministerie van Volksgezondheid en Milieuhygiëne, 22

GILBERT, O.L. (1965) Lichens as indicators of air pollution in the Tyne Valley In: Goodman, G.T., Edwards, R.W. and Lambert, J.M. (eds.) Ecology and the industrial society, 35-47 Oxford, Blackwell.

GILBERT, O.L. (1968) Biological indicators of air pollution. Ph.D. thesis University of Newcastle upon Tyne.

GILBERT, O.L. (1973) The Effect of airborne fluorides. In: Ferry, B.W., Baddeley, M.S. & Hawksworth, D.L. (eds.) Air Pollution and Lichens 167-91. London, The Athlone Press.

GILBERT, O.L. (1974) An air pollution survey by school children. Environ. Pollut. 6, 175-80.

GRODZINSKA, K. (1971) Acidification of tree bark as a measure of air pollution in southern Poland. Bull. Acad. Polon. Sci., ser.sci. biol. II, 19-189-95.

HAWKSWORTH, D.L. (1971) Lichens as a litmus for air pollution: A historical review Int. J. Environ. Studies, 1, 281-96.

HAWKSWORTH, D.L. (1973) Mapping StudiesIn: Ferry, B.W., Baddeley, M.S. and Hawksworth, D.L. (eds.) Air Pollution and Lichens., 38-76. London, The Athlone Press.

HAWKSWORTH, D.L. AND ROSE, F. (1970) Qualitative scale for estimating sulphur dioxide air pollution in England and Wales using epiphytic lichens. Nature, Lond. 227, 145-8.

HAWKSWORTH, D.L., ROSE, F. AND COPPINS, B.J. (1973) Changes in the Lichen Flora of England and Wales Attributable to Pollution of the Air by Sulphur Dioxide In: Ferry, B.W., Baddeley, M.S. and Hawksworth, D.L. (eds.) Air Pollution and Lichens, 330-67. London, The Athlone Press.

HILL, D.J. (1974) Some effects of sulphite on photosynthesis in lichens New Phytol. 73, 1193-1205.

HORNTVEDT, R. (1975) Epiphytic macrolichens on Scots pine related to air pollution from industry in Odda, Western Norway. Rep. Norw. Forest Res. Inst. 31-12.

JAMES, P.W. (1965) A new checklist of British lichens. Lichenologist, 3,95-153.

JAMES, P.W. (1973) The effect of air pollutants other than hydrogen fluoride and sulphur dioxide on lichens. In: Ferry, B.W., Baddeley, M.S. & Hawksworth, D.L. (eds.) Air Pollution and Lichens. 143-75. London, The Athlone Press.

JAMES, P.W. AND ROSE, F. (1974) Map 15. Pertusaria coccodes. Lichenologist 6,191-3.

JÜRGING, P. (1975) Epiphytische Flechten als Bioindikatoren der Luftverunreinigung. Vaduz, Cramer. Bibliotheca Lichenologica 4.

KUNZE, M. (1974) Flechten als Indikatoren von Luftverunreinigungen. Beihefte zu den Veröffentlichungen der Landesstelle für Naturschutz und Landschaftspflege Baden-Württemberg 5.

K.N.M.I. (1972) Klimaatatlas van Nederland 's-Gravenhage: Staatsuitgeverij.

LEBLANC, F. (1971) Possibilities and methods for mapping air pollution on the basis of lichen sensitivity. Mitt.forstl. Bund Vers Anst. Wien 92,103-26.

LEBLANC, F. & RAO, D.N. (1966) Réaction de quelques lichens et mousses epiphytiques à l'anhydride sulfureux dans la région de Sudbury, Ontario. Bryologist, 69, 338-46.

LEBLANC, F. and DE SLOOVER, J. (1970) Relation between industrialization and the distribution and growth of epiphytic lichens and mosses in Montreal. Can. J. Bot. 48, 1485-96.

LEBLANC, F., COMEAU, G. & RAO, D.N. (1971) Fluoride injury symptoms in epiphytic lichens and mosses. Can. J. Bot. 49, 1691-8.

LEBLANC, F. & RAO, D.N. (1973a) Evaluation of the pollution and drought hypotheses in relation to lichens and bryophytes in urban environments. Bryologist 76, 1-19.

LEBLANC, F. & RAO, D.N. (1973b) Effects of sulphur dioxide on lichen and moss transplants. Ecology, 54, 612-7.

MARGOT, J. (1973) Experimental study of the effects of sulphur dioxide on the soredia of Hypogymnia physodes In: Ferry, B.W., Baddeley, M.S. & Hawksworth, D.L. (eds.) Air Pollution and Lichens. 314-29. London, The Athlone Press.

MELTZER, J.A. (1976) Epiphytische Lichenen en Luchtverontreiniging in Noord-Holland, (not published).

MOOI, J. (1974) Onderzoek naar de gevoeligheid en opnamecapaciteit van houtige gewassen voor waterstoffluoride (HF) Groen, 10,321-6.

MOORE, C.C. (1974) A Modification of the 'Index of Atmospheric Purity' method for substrate differences. Lichenologist 6,156-7.

NASH, Thomas H. (1971) Lichen sensitivity to hydrogen fluoride. Bull. Torrey Bot.Club 98,103-6.

NASH, Thomas H. (1973) Sensitivity of lichens to sulphur dioxide. Bryologist, 76,333-9.

PEARSON, L.C. & SKYE, E. (1965) Air pollution effects pattern of photosynthesis in Parmelia sulcata, a corticolous lichen. Science, N.Y. 148,1600-2.

PEARSON, L.C. (1973) Air pollution and lichen physiology: progress and problems In: Ferry, B.W., Baddeley, M.S. & Hawksworth, D.L. (eds.) Air Pollution and Lichens. 224-237. London, The Athlone Press.

PEN, J. (1970) Harmonie en conflict. Amsterdam: De Bezige Bij.

PUCKETT, K.J., NIEBOER, E., FLORA, W.P. & RICHARDSON, D.H.S. (1973) Sulphur dioxide: its effect on photosynthetic ^{14}C fixation in lichens and suggested mechanisms of phytotoxity. New Phytol. 72,141-154.

PUCKETT, K.J., RICHARDSON, D.H.S., FLORA, W.P. & NIEBOER, E. (1974) Photosynthetic ^{14}C fixation by the lichen Umbilacaria muhlenbergii (Ach.) Tuck. following short exposures to acqueous sulphur dioxide New Phytol. 73,1183-92.

PYATT, F.B. (1970) Lichens as indicators of air pollution in a steel producing town in South Wales. Environ.Pollut. 1,45-56.

RAO, D.N. AND LEBLANC, F. (1965) Effects of sulphur dioxide on the lichen algae with special reference to chlorophyll. Bryologist 69,69-75.

RICHARDSON, D.H.S. & PUCKETT, K.J. (1973) Sulphur Dioxide and Photosynthesis in Lichens. In: Ferry, B.W., Baddeley, M.S. & Hawksworth D.L. (eds.) Air Pollution and Lichens. 283-99. London, The Athlone Press.

SCHÖNBECK, H. (1969) Eine Methode zur Erfassung der biologischen Wirkung von Luftverunreinigungen durch transplantierte Flechten. Staub-Reinhaltung der Luft 29,1,14-8.

SCHÖNBECK, H. (1972) Untersuchungen in Nordrhein-Westfalen über Flechten als Indikatoren für Luftverunreinigungen. Schriftenreihe der LIB des Landes NW 26,99-104.

SKYE, E. (1968) Lichens and air pollution. Acta phytogeogr. suec. 52, 1-123.

SOKAL, R.R. & ROHLF, F.J. (1969) Biometry. San Francisco, Freeman and company.

SPIERING, F. (1974) Onderzoekingen over de invloed van HF luchtverontreiniging op planten. Landbouwkundig Tijdschrift 86-12, 331-5.

TRASS, H. (1968) Indeks samblikur ühmituste kasutamiscks õhu saastatuse määramisel. Eesti Loodus 11, 628.

TRASS, H. (1968) Samblikud-õhu saastatuse inditaatorid Eesti Loodus 11, 80-3.

TRASS, H. (1973) Lichen sensitivity to the air pollution and index of poleotolerance (I.P.) Fol. Crypt Est. Fasc. 3 p. 17-24 Tartu 1973.

TÜRK, R., WIRTH, V. & LANGE, O.L. (1974) CO_2-Gaswechsel-Untersuchungen zur SO_2-Resistenz von Flechten. Oecologia (Berl.) 15, 33-64.

TÜRK, R. & WIRTH, V. (1975) The pH Dependence of SO_2 Damage to Lichens Oecologia (Berl.) 19,285-91.

WIRTH, W. & TÜRK, R. (1975) Zur SO_2-Resistenz von Flechten verschiedener Wuchsform. Flora, 164, 133-43.

appendix 1

EPIPHYTIC VEGETATION IN THE NETHERLANDS

Situation ca. 1950
after: Barkman 1969

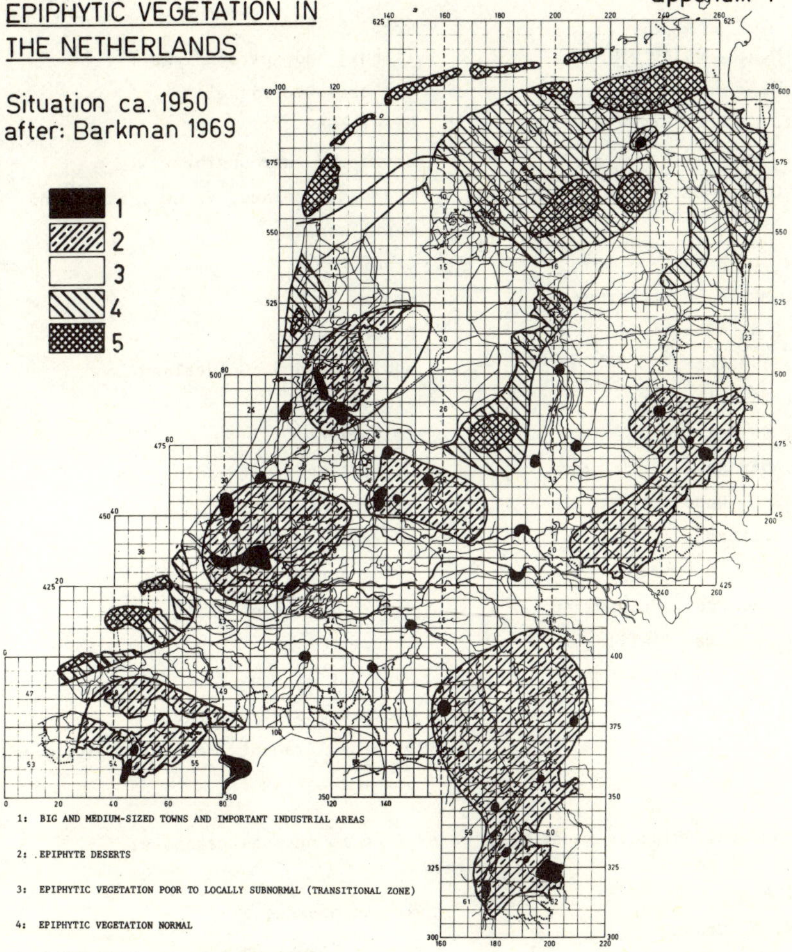

1: BIG AND MEDIUM-SIZED TOWNS AND IMPORTANT INDUSTRIAL AREAS

2: EPIPHYTE DESERTS

3: EPIPHYTIC VEGETATION POOR TO LOCALLY SUBNORMAL (TRANSITIONAL ZONE)

4: EPIPHYTIC VEGETATION NORMAL

5: EPIPHYTIC VEGETATION RICH AND LUXURIANT

List of WHEN-participants and main area of fieldwork:

M. Brand	Zuid-Holland, Zeeland, Overijssel
H.F. v. Dobben	Brabant, Zeeland, Veluwe, Overijssel, Zuid-Holland
R.M.L.A. Dolné	Tilburg and surroundings
H.W.J. v. Dijk	Limburg, Overijssel
Ms. C. Leemburg-van der Graaf	Groningen
B. Husson	Breda and surroundings
W.O. van der Knaap	Brabant, Zeeland, Veluwe, Friesland, Overijssel
Ms. L. Kraak	Groningen
Kromme Rijn Werkgroep	Kromme Rijn-area
Ms. B. Mansell	Gelderland
J.A. Meltzer	Noord-Holland, Gelderland
P. Muizelaar	Friesland
H. de Neeling	Gelderland
H. Sipman	Utrecht and surroundings
Ms. H. Uppelschoten-Sliepen	Utrecht and surroundings
K. Slings	Noord-Holland
J.C. Smittenberg	Groningen
Ms. R. Swart-Velthuizen	Environs of Bennekom
Mr. and Ms. Tjallingii	Betuwe
Ms. S. van der Vaart-de Vlieger	Environs of Bennekom
K. Westerbrink	Noord-Holland, Utrecht
Ms. A. de Wit	Drenthe, Brabant, Zeeland, Gelderland, Utrecht

Other participants whose data were given to one of the above:

F. Bos
Ms. N. Drost
H. During
J.B.M. Frencken
Ms. D. Brand-de Heer
W. de Kleine
J. Lankelma
B. Maas
Ms. A. Meiners
M. van Noordwijk
Ms. W. Riemersma
M. de Schiffart
Ms. N. Schmidt
Ms. E. Smit
J. Steenhuis
W. Wouters

List of species to be mapped according to the initial WHEN project design

	sensitivity values (Q) according to Barkman
Protococcus viridis	0
Lecanora conizaeoides (sterile)	0
Lecanora conizaeoides (fertile)	1
Dicranoweisia cirrhata	2
Prasiola crispa	2
Lepraria incana	2
Buellia punctata	3
Lecidea scalaris	3
Parmelia physodes	4
Buellia canescens	4
Lecanora chlarotera s.l.	4
Physcia grisea	5
Physcia adscendens + Physcia tenella	5
Parmelia subrudecta	6
Parmelia sulcata	6
Pertusaria albescens	6
Xanthoria parietina	7
Parmelia acetabulum	7
Ramalina farinacea	7
Homalotheciun sericeum	7
Metzgeria furcata	8
Parmelia caperata	8
Evernia prunastri	8
Parmelia furfuracea	8
Ramalina fastigiata	8
Pertusaria amara	8
Lecidea limitata	8
Graphis scripta	9
Physcia pulverulenta	9
Physcia aipolia	9
Cetraria chlorophylla	9
Leucodon sciuroides	9
Frullania dilatata	9
Parmelia tiliacea	9
Cetraria glauca	10
Anaptychia ciliaris	10
Ramalina fraxinea	10
Parmelia saxatilis	10
Usnea spp.	12

List of excluded taxa:

Arthonia

Arthopyrenia

Bacidia

Caloplaca

Catillaria prasina

Cladonia

Dimerella

Gyalecta

Lecania

Lecidea granulosa

Lecidea uliginosa

Ochrolechia

Porina

Pyrenula

Rinodina

Xanthoria aureola is combined with X. parietina under the latter name

For Pertusaria albescens no varieties are distinguished.

Physcia grisea sl. = Physcia grisea + Physcia farrea.

Physcia adscendens and Physcia tenella are taken together as the latter.

All Usnea species are combined as one taxon Usnea spp.

All species of the Lecanora subfusca group are combined under the name Lecanora subfusca s.l.

appendix 5

Cumulative frequency distribution of the number of lichen species per tree species

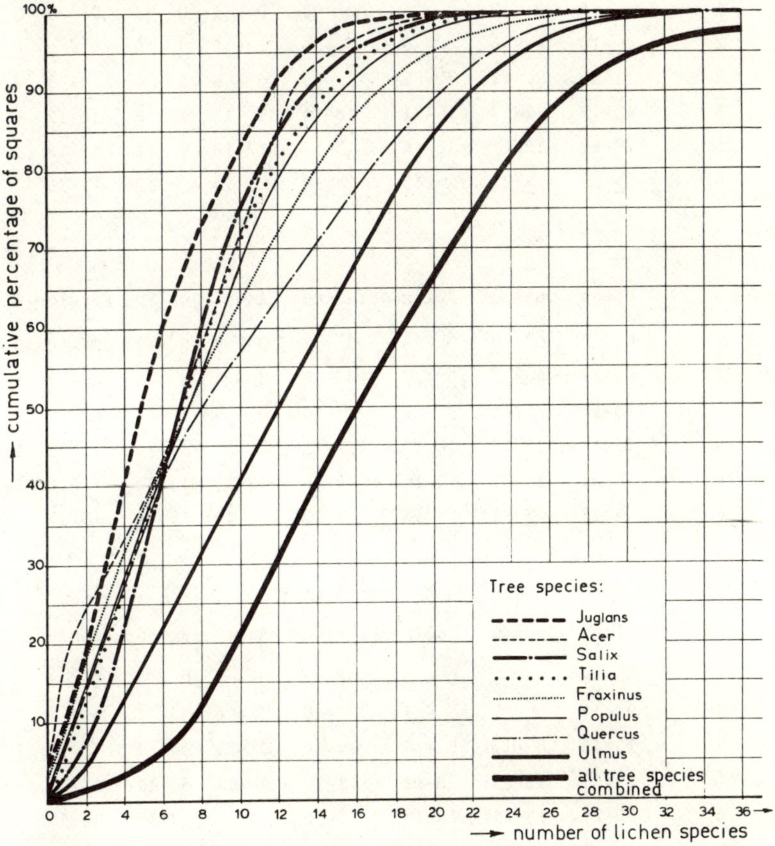

Interpretation: for example, in 20% of all squares with Ulmus, this tree species has less than 6 lichen species.

Appendix 6

Numbers of lichens per class per tree species.

Initial classification T_0

epiphyte class	Ulmus	Quercus	Salix	Fraxinus	Acer	Juglans	Populus	Tilia
1	0 - 5	0 - 3	0 - 4	0 - 2	0 - 1	0 - 2	0 - 3	0 - 3
2	6 -10	4 - 6	5 - 6	3 - 6	2 - 5	3 - 4	4 - 6	4 - 6
3	11-14	7 -11	7 - 8	7 - 9	6 - 8	5 - 6	7 - 9	7 - 8
4	15-19	12-17	9 -11	10-14	9 -11	7 - 8	10-12	9 -12
5	20-22	18-21	12-14	15-17	12-13	9 -12	13-15	13-15
6	$\geq$ 23	$\geq$ 22	$\geq$ 15	$\geq$ 18	$\geq$ 14	$\geq$ 13	$\geq$ 16	$\geq$ 16

T_1

	Ulmus	Quercus	Salix	Fraxinus	Acer	Juglans	Populus	Tilia
1	0 - 7	0 - 3	0 - 4	0 - 2	0 - 2	0 - 2	0 - 3	0 - 4
2	8 -12	4 - 7	5 - 7	3 - 6	3 - 7	3 - 4	4 - 7	5 - 7
3	13-16	8 -12	8 - 9	7 -10	8 - 9	5 - 7	8 -10	8 -10
4	17-19	13-17	10-13	11-13	10-11	8 -11	11-13	11-13
5	20-22	18-21	14-15	14-16	12-13	12-13	14-16	14-15
6	$\geq$ 23	$\geq$ 22	$\geq$ 16	$\geq$ 17	$\geq$ 14	$\geq$ 14	$\geq$ 17	$\geq$ 16

final classification $T_4 = T_5 =$ ....

	Ulmus	Quercus	Salix	Fraxinus	Acer	Juglans	Populus	Tilia
1	0 - 8	0 - 3	0 - 4	0 - 2	0 - 2	0 - 2	0 - 3	0 - 3
2	9 -12	4 - 7	5 - 7	3 - 6	3 - 6	3 - 4	4 - 6	4 - 6
3	13-15	8 -12	8 - 9	7 -10	7 - 9	5 - 7	7 -10	7 - 9
4	16-19	13-17	10-13	11-13	10-11	8 -11	11-13	10-13
5	20-22	18-21	14-15	14-16	12	12-13	14-16	14-15
6	$\geq$ 23	$\geq$ 22	$\geq$ 16	$\geq$ 17	$\geq$ 13	$\geq$ 14	$\geq$ 17	$\geq$ 16

appendix 7

COMBINED MAP OF THE EPIPHYTIC LICHEN VEGETATION BASED ON INITIALCLASSIFICATION T_O (see also app. 6)

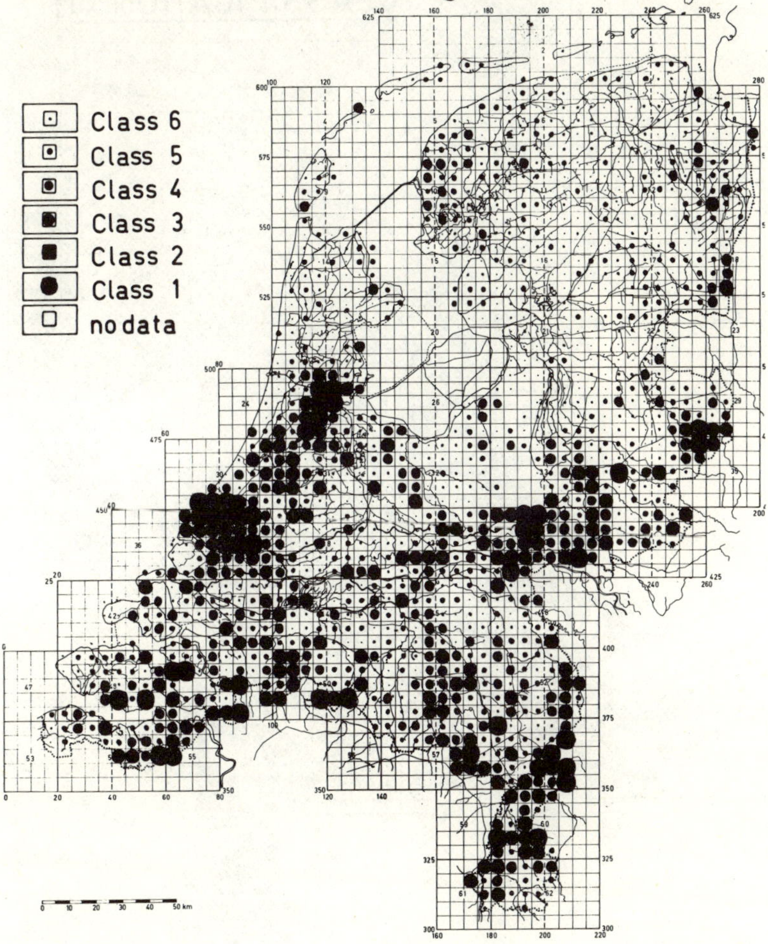

appendix 8

COMBINED MAP OF THE EPIPHYTIC LICHEN VEGETATION BASED ON CLASSIFICATION T_1

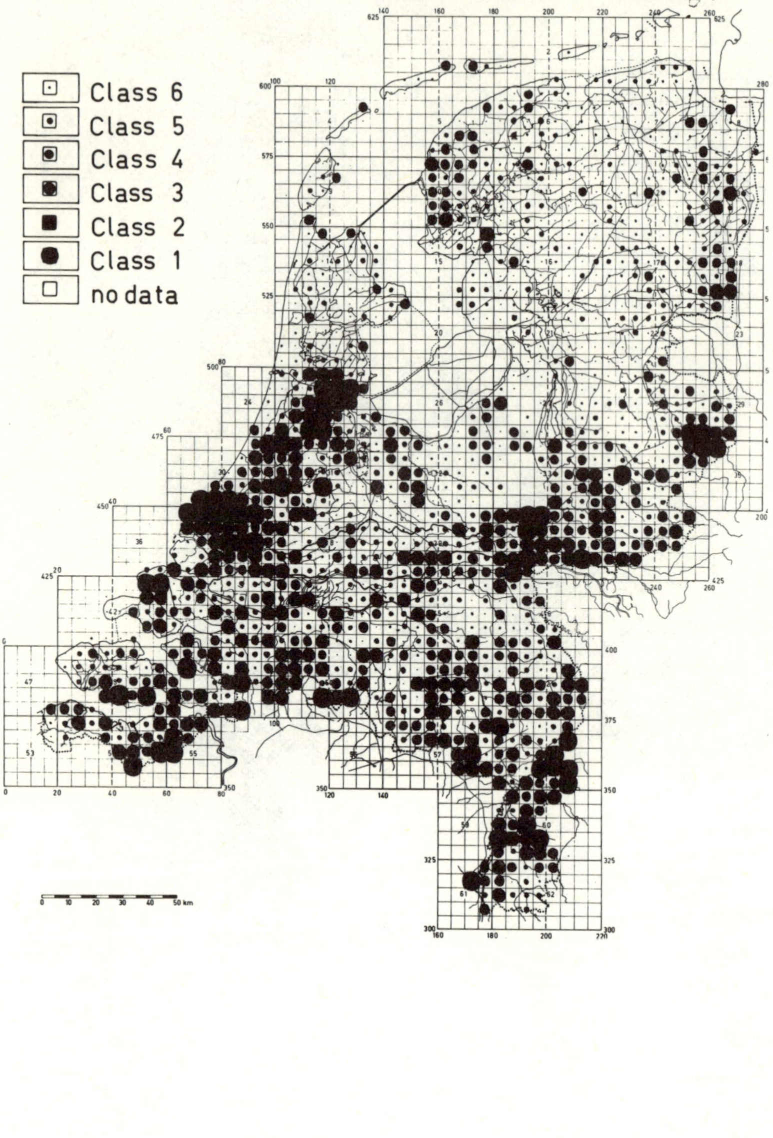

appendix 14

POPULUS

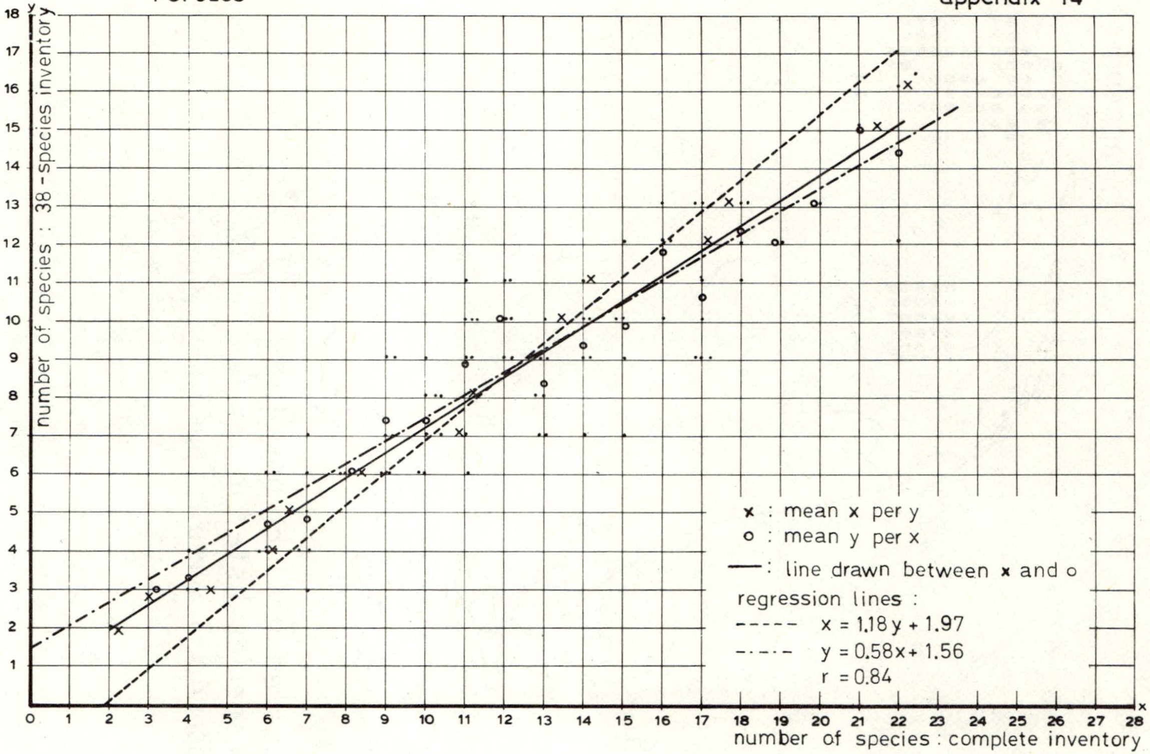

appendix 13

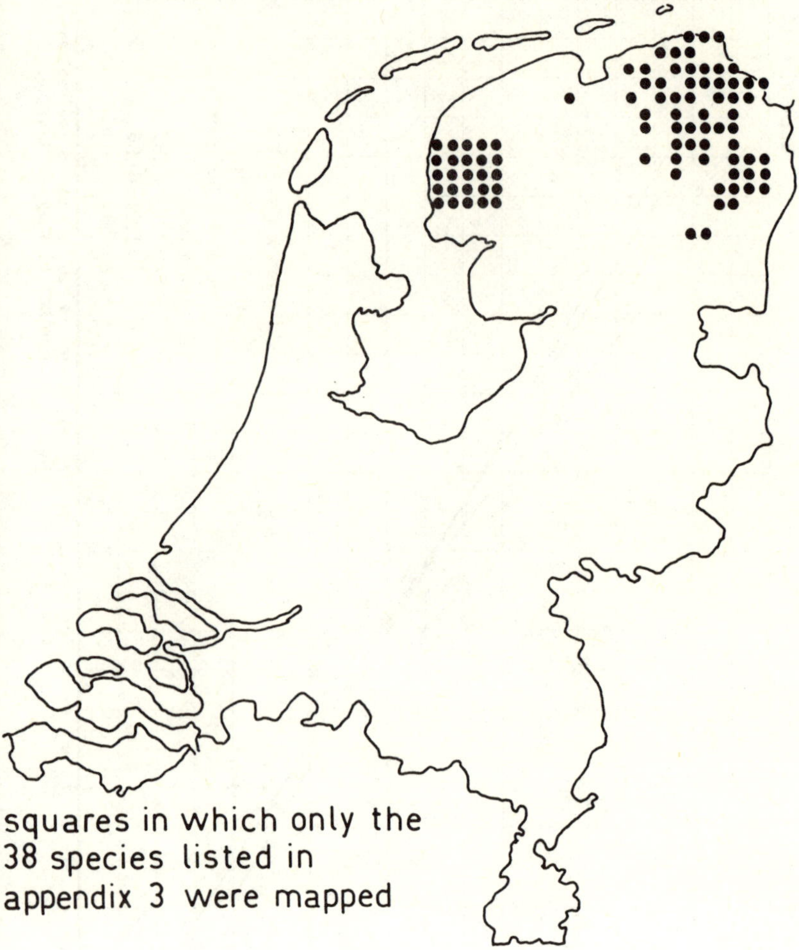

squares in which only the 38 species listed in appendix 3 were mapped

appendix 15

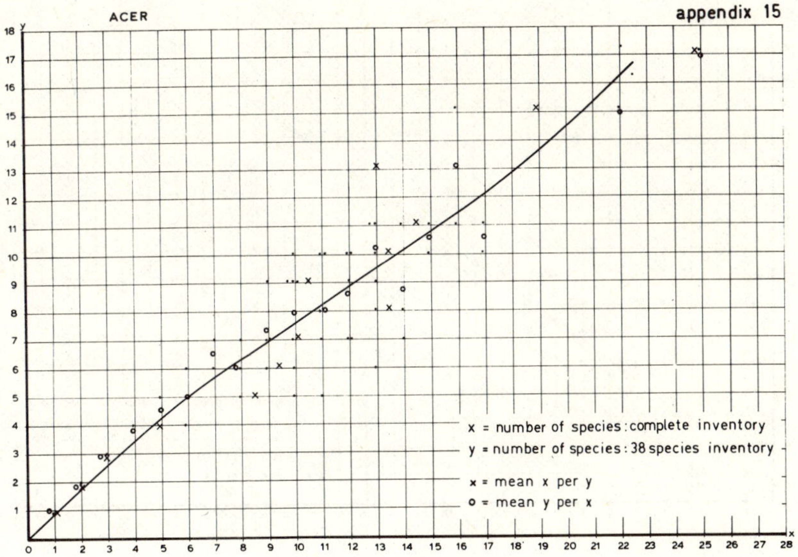

appendix 16

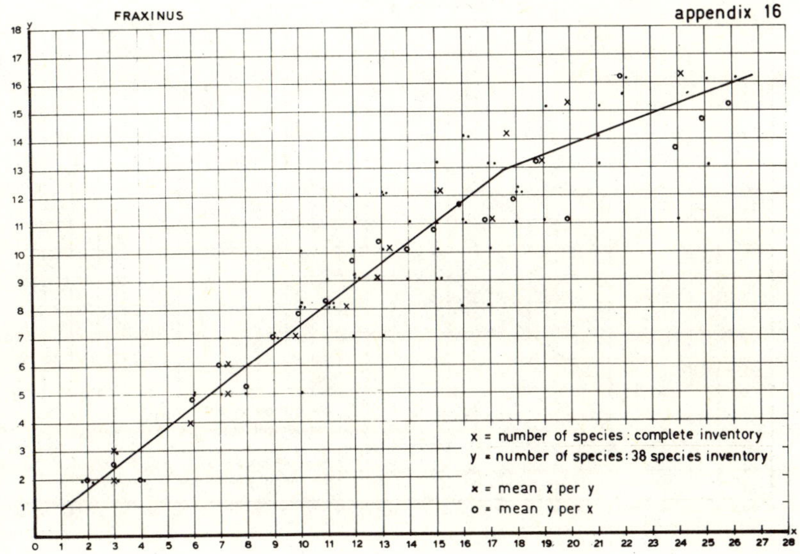

appendix 17

QUERCUS

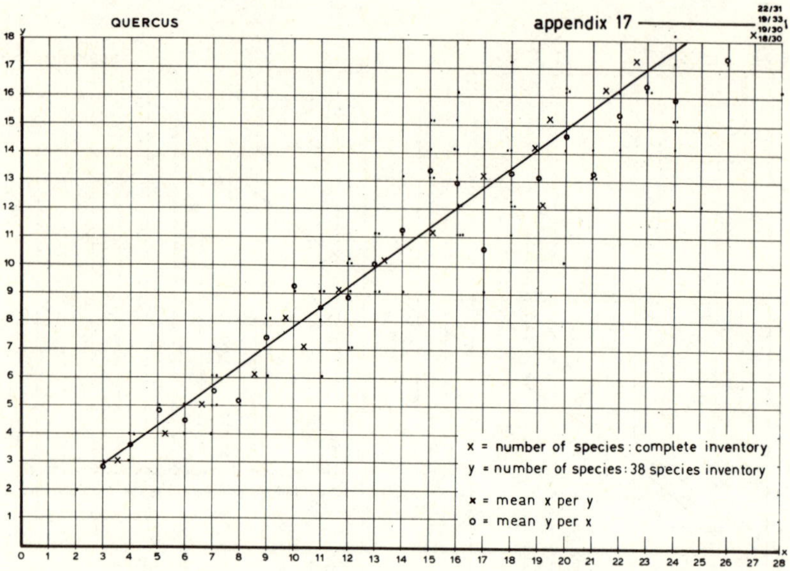

appendix 18

ULMUS

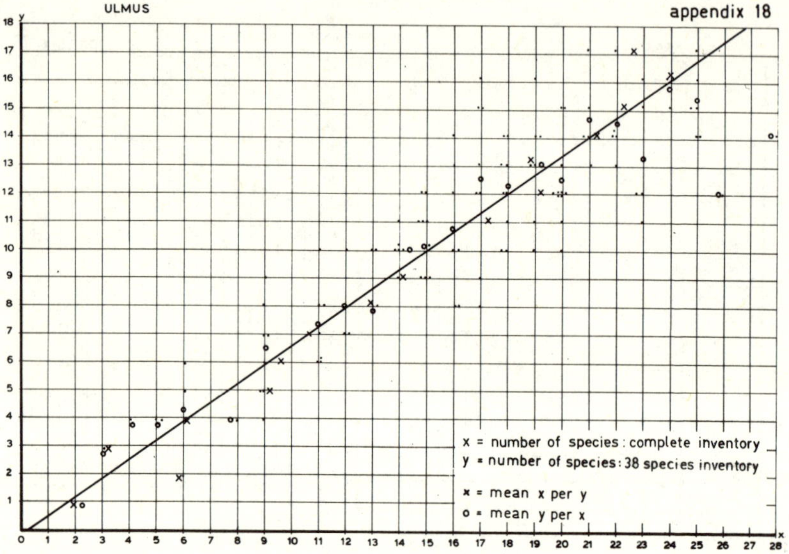

appendix 19

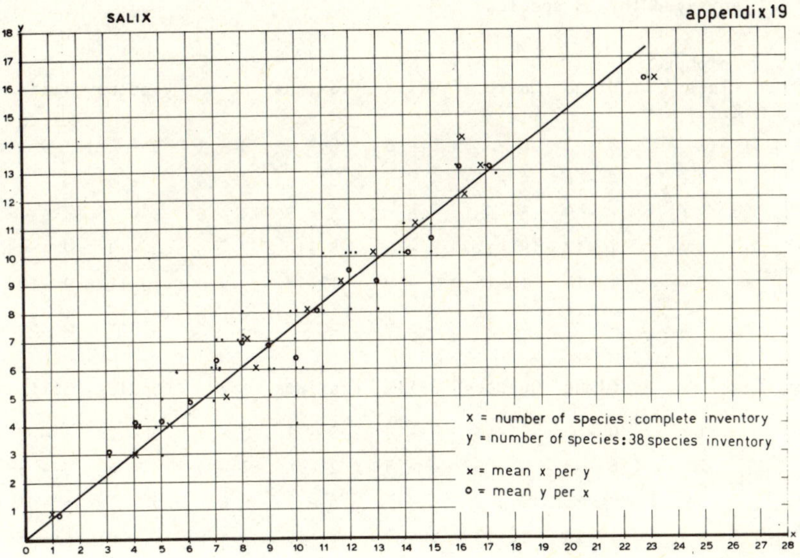

appendix 20

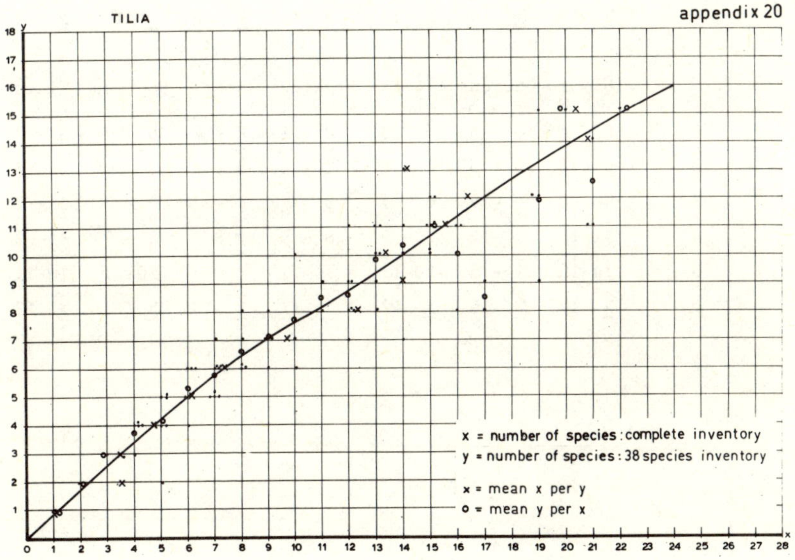

Number of lichens per class per tree species for squares surveyed for 38 species.

epiphyte class T_0'	Ulmus	Quercus	Salix	Fraxinus	Acer	Populus	Tilia
1	0 - 3	0 - 3	0 - 3	0 - 2	0 - 1	0 - 2	1 - 2
2	4 - 6	4 - 5	4 - 5	3 - 4	2 - 4	3 - 5	3 - 5
3	7 - 9	6 - 8	6	5 - 7	5 - 6	6 - 7	6
4	10-12	9 -12	7 - 8	8 -10	7 - 8	8 - 9	7 - 8
5	13-14	13-15	9 -10	11-12	9	10-11	9 -10
6	$\geq$ 15	$\geq$ 16	$\geq$ 11	$\geq$ 13	$\geq$ 10	$\geq$ 12	$\geq$ 11

T_1'	Ulmus	Quercus	Salix	Fraxinus	Acer	Populus	Tilia
1	0 - 5	0 - 3	0 - 3	0 - 1	0 - 1	0 - 2	1 - 2
2	6 - 8	4 - 6	4 - 5	2 - 4	2 - 5	3 - 4	3 - 5
3	9 -11	7 - 9	6 - 7	5 - 7	6	5 - 7	6 - 7
4	12-13	10-13	8 - 9	8 - 9	7 - 8	8 - 9	8 - 9
5	14-15	14-15	10-11	10-11	9	10-11	10-11
6	$\geq$ 16	$\geq$ 16	$\geq$ 12	$\geq$ 12	$\geq$ 10	$\geq$ 12	$\geq$ 12

T_4'	Ulmus	Quercus	Salix	Fraxinus	Acer	Populus	Tilia
1	0 - 5	0 - 3	0 - 3	0 - 1	0 - 2	0 - 2	0 - 2
2	6 - 8	4 - 5	4 - 5	2 - 4	3 - 5	3 - 4	3 - 5
3	9 -10	6 - 9	6 - 7	5 - 7	6 - 7	5 - 7	6 - 7
4	11-13	10-13	8 - 9	8 - 9	8	8 - 9	8 - 9
5	14-15	14-15	10-11	10-11	9	10-11	10-11
6	$\geq$ 16	$\geq$ 16	$\geq$ 12	$\geq$ 12	$\geq$ 10	$\geq$ 12	$\geq$ 12

appendix 22

ACER

⊡	Class 6
⊡	Class 5
⊡	Class 4
⊡	Class 3
■	Class 2
●	Class 1

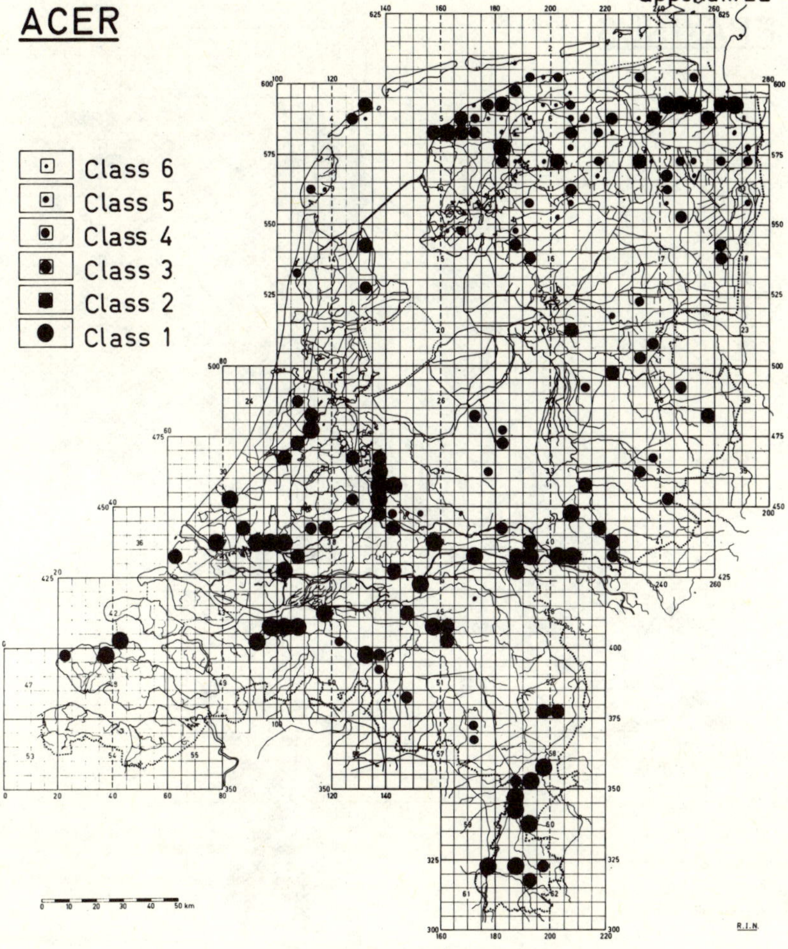

FRAXINUS

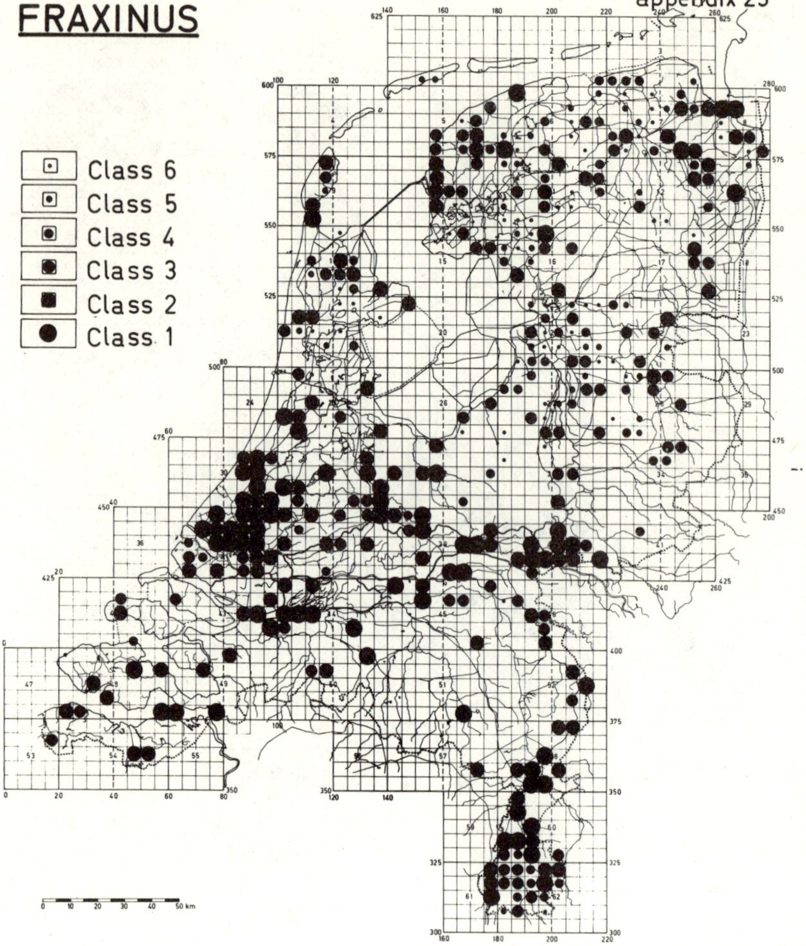

appendix 24

JUGLANS

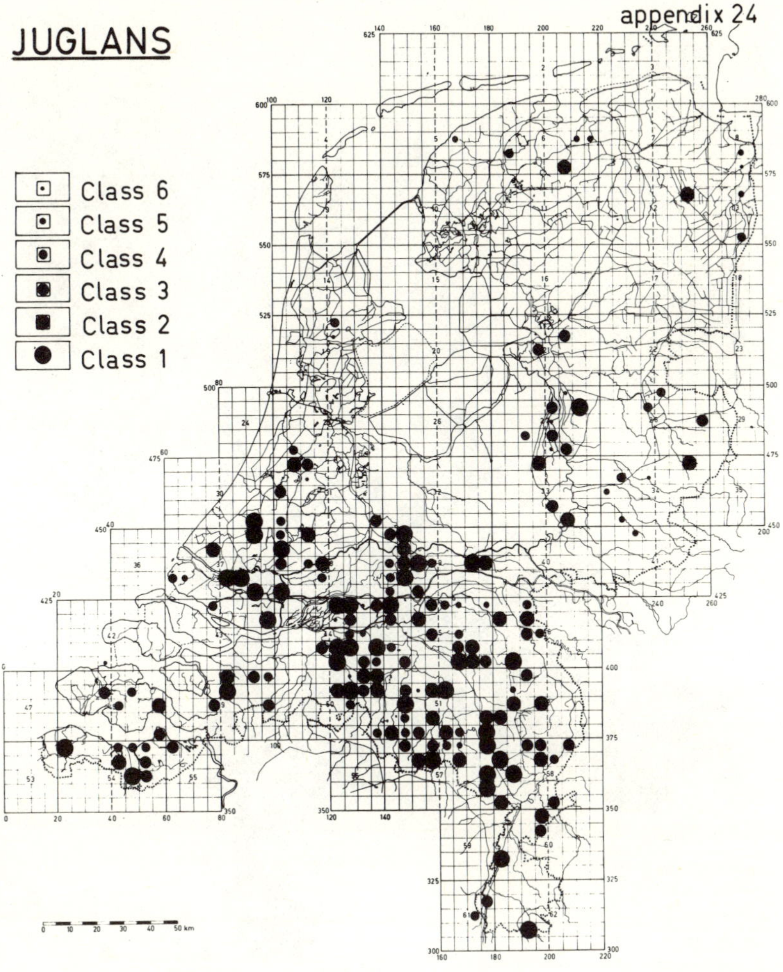

appendix 25

POPULUS

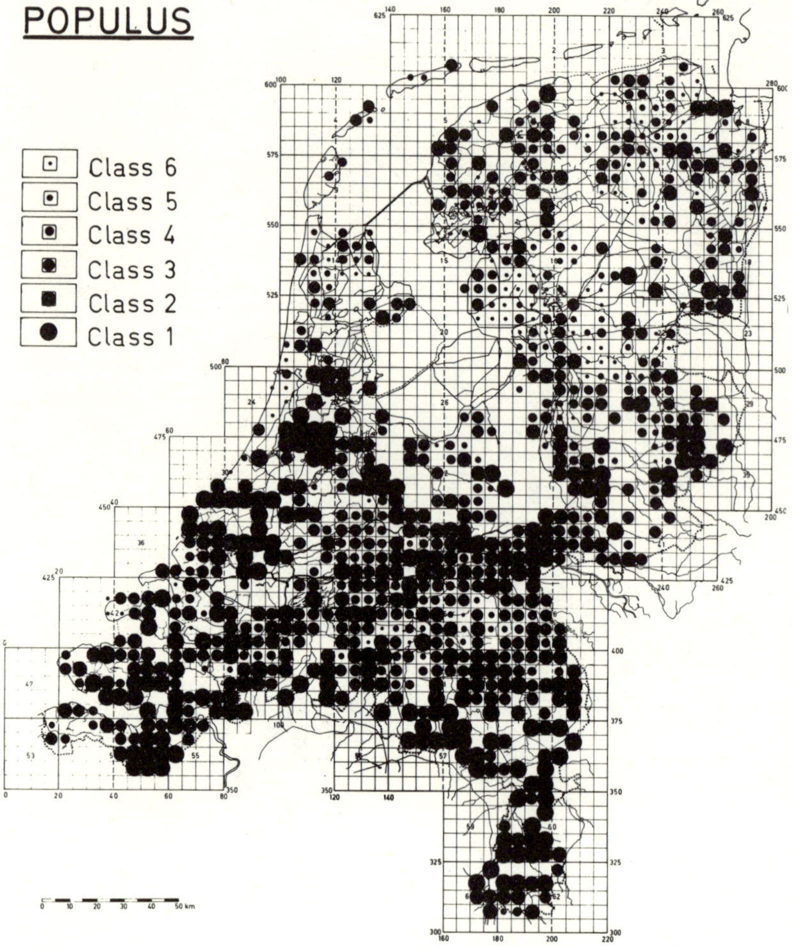

appendix 26

QUERCUS

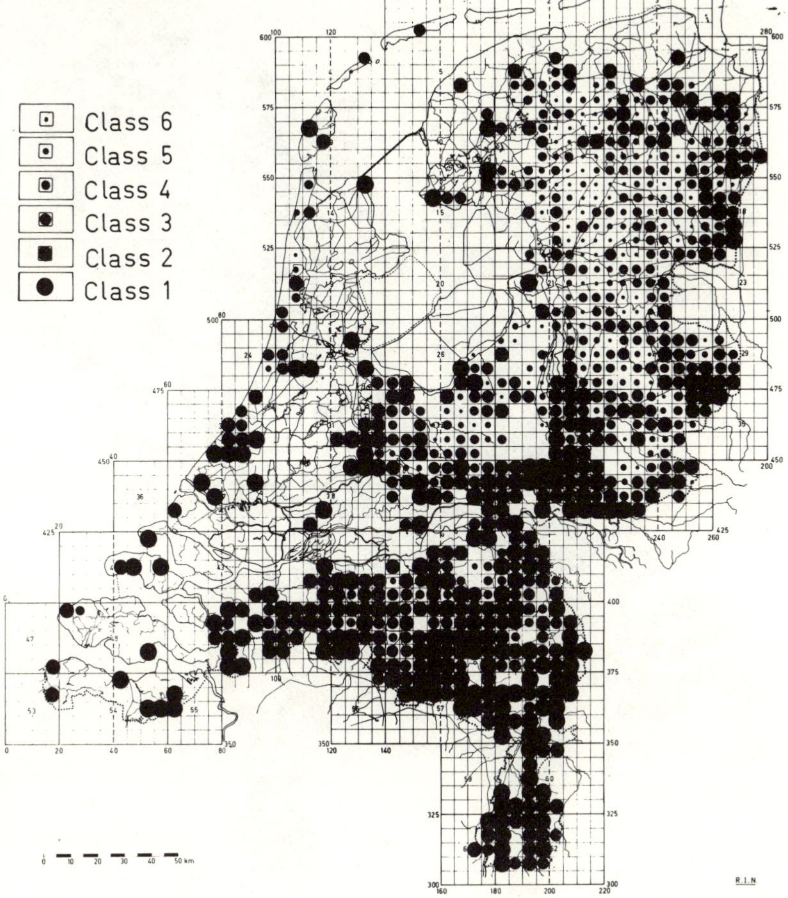

appendix 27

SALIX

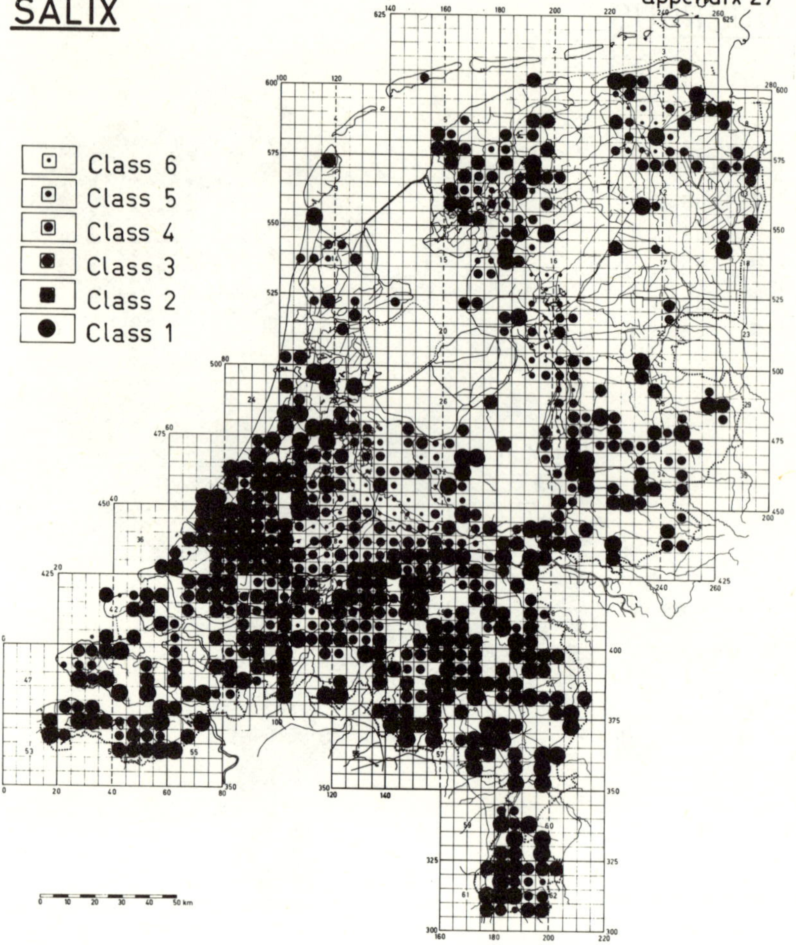

appendix 28

TILIA

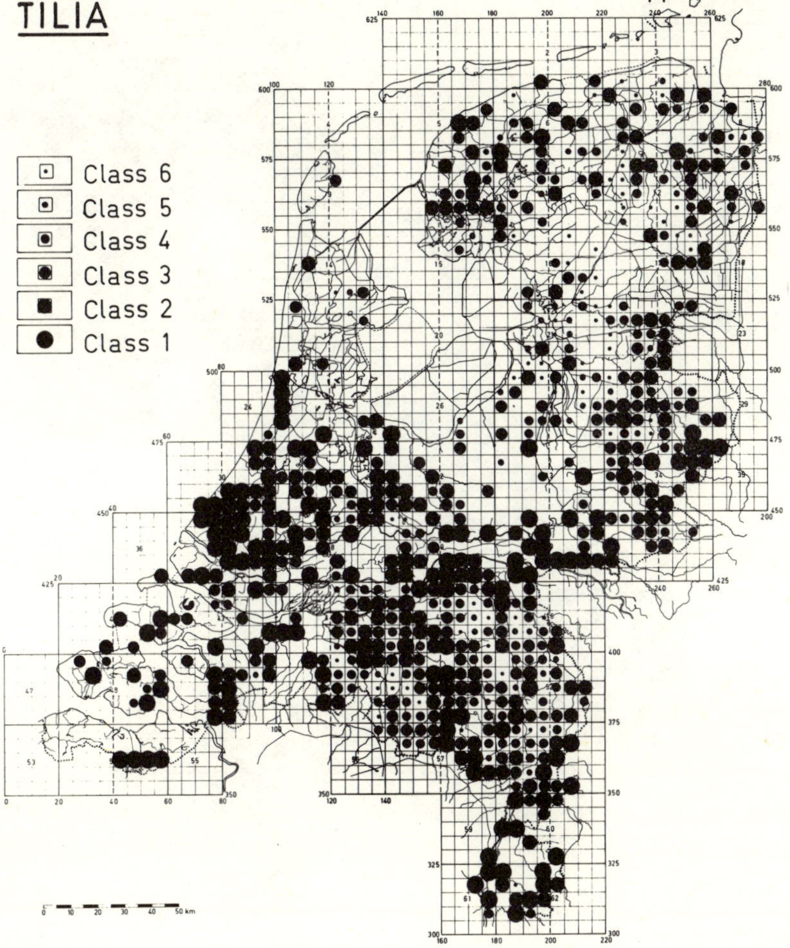

appendix 29

ULMUS

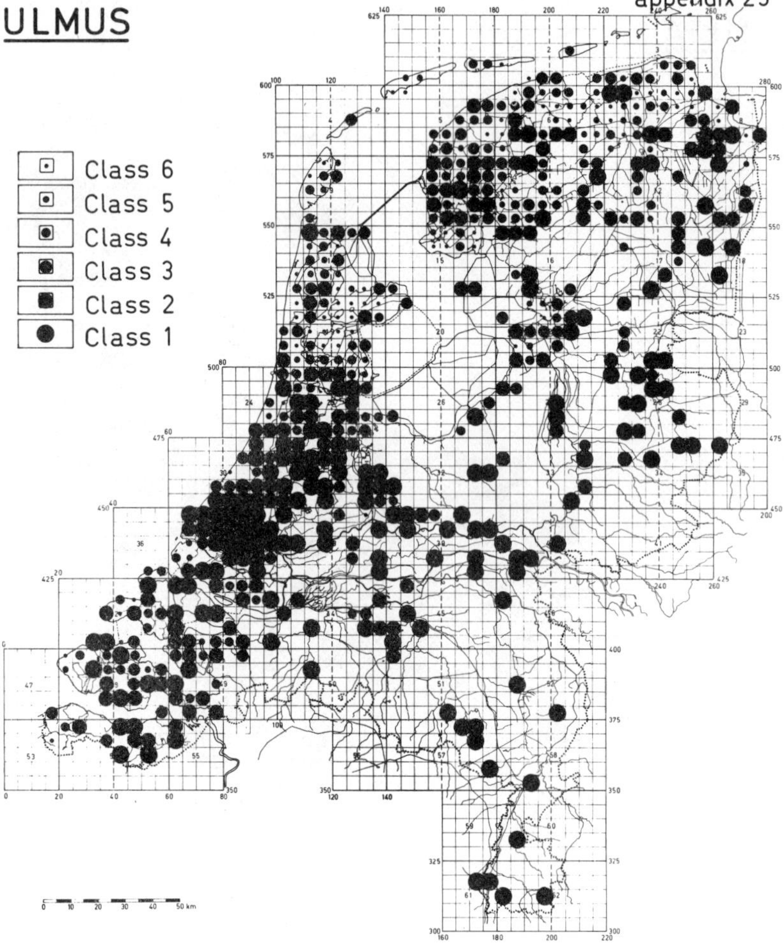

appendix 30

GENERALIZED MAP OF EPIPHYTIC LICHEN VEGETATION IN THE NETHERLANDS

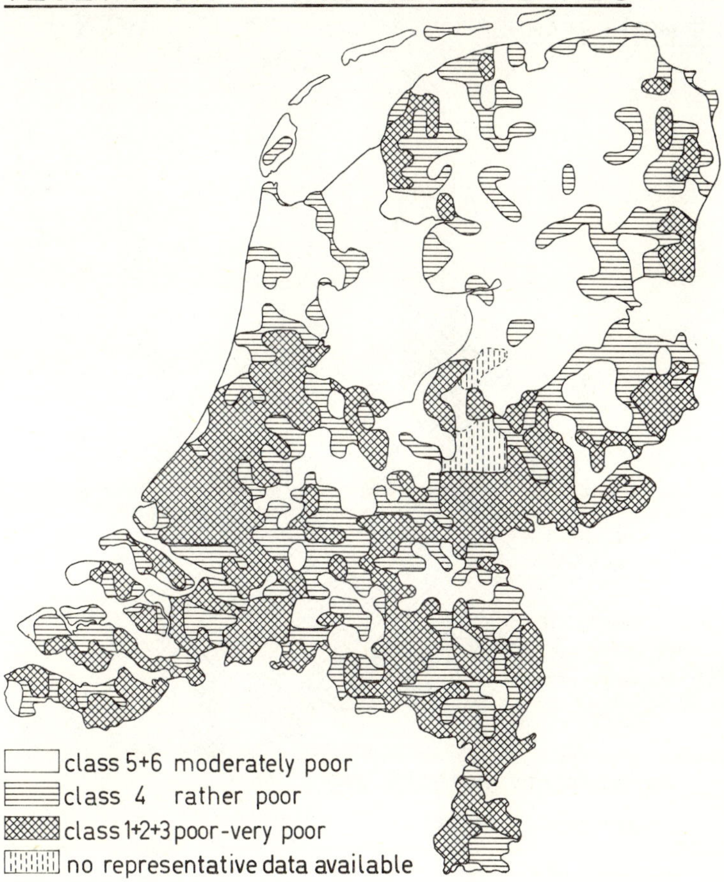

appendix 31

ROUGH MAP OF EPIPHYTIC LICHEN VEGETATION IN THE NETHERLANDS

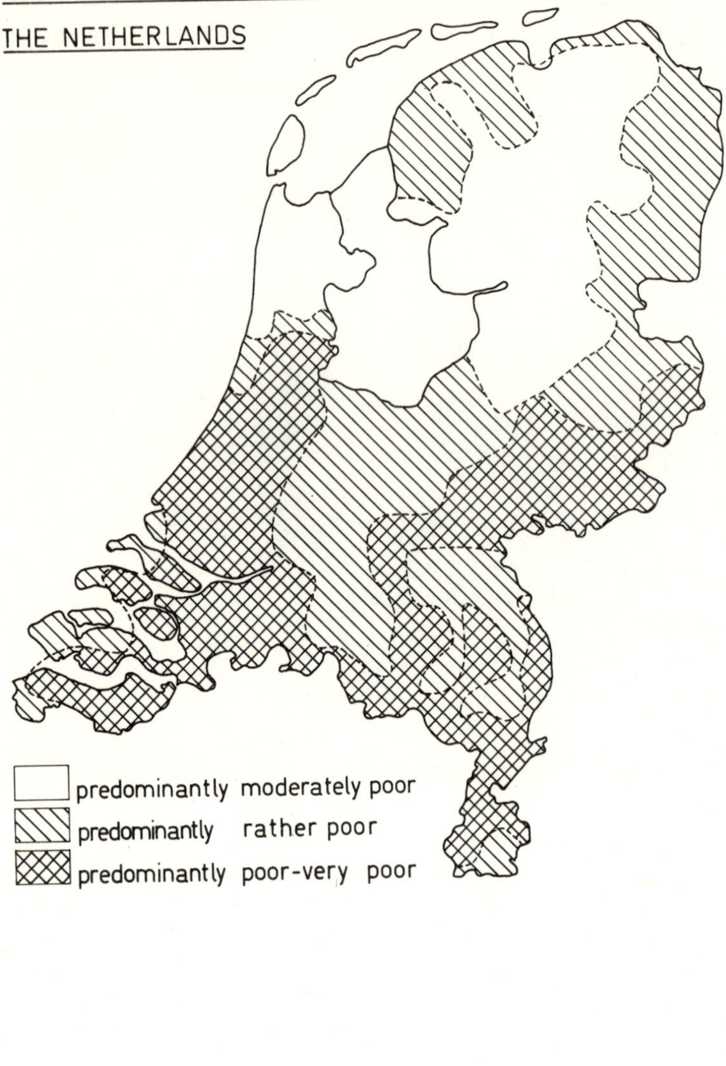

appendix 32

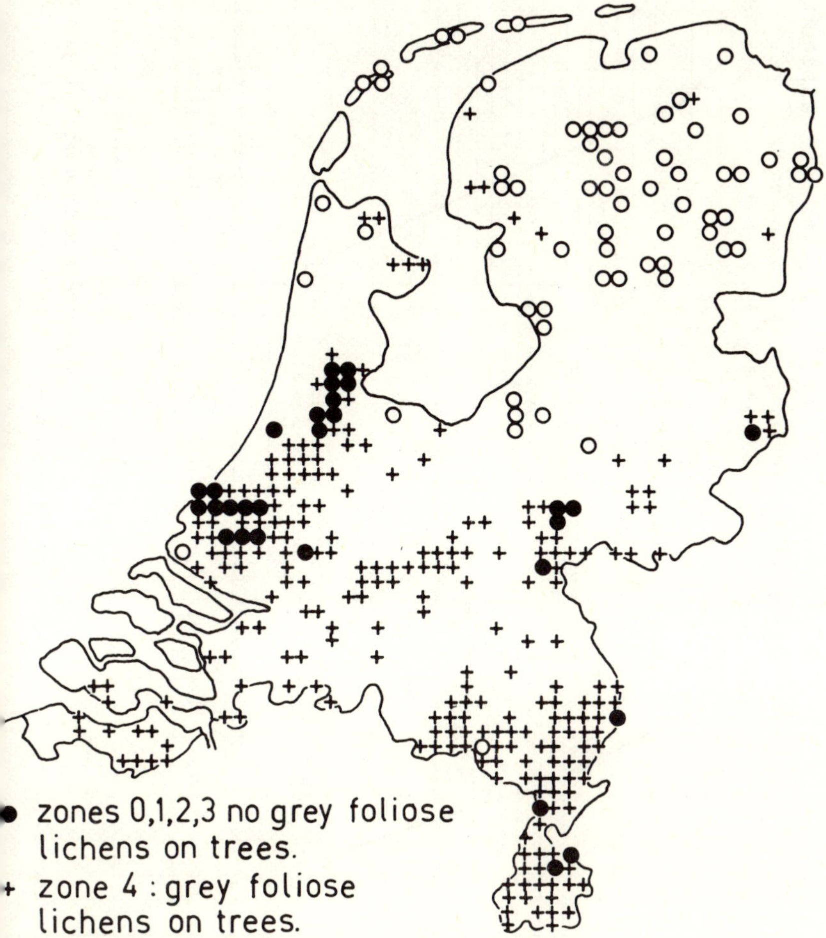

● zones 0,1,2,3 no grey foliose lichens on trees.

+ zone 4 : grey foliose lichens on trees.

zone 5 : fruticose lichens on trees.

ο zone 6 : Usnea

zones as in Gilbert 1974

appendix 33

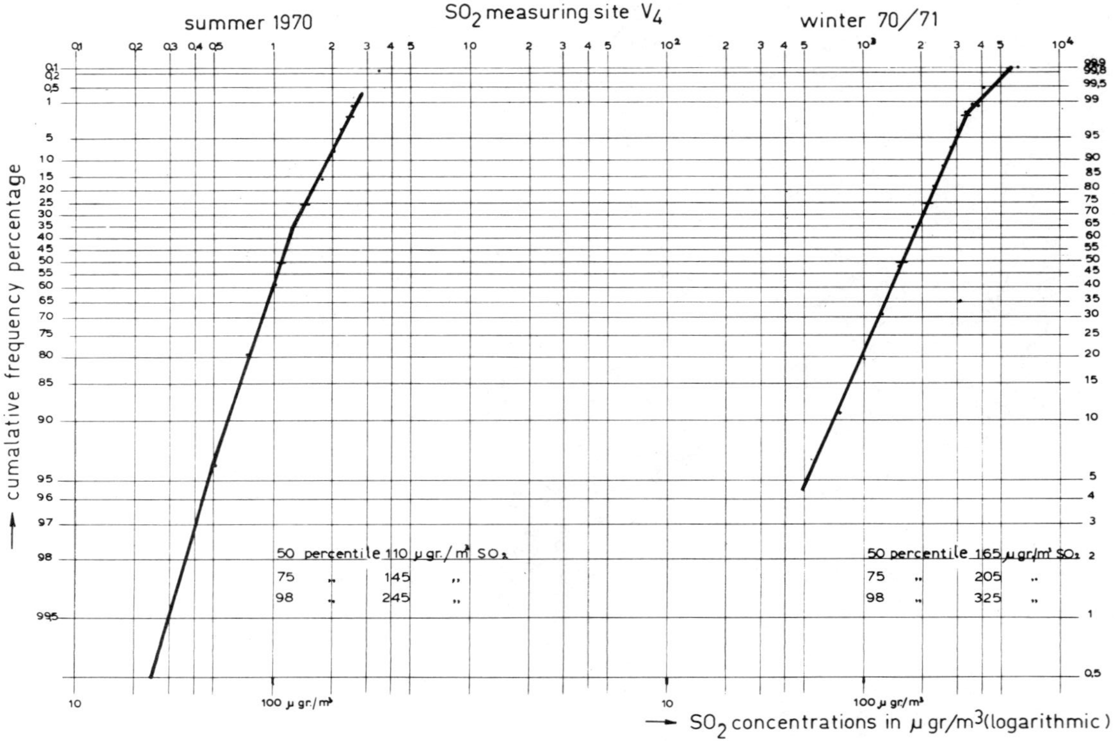

example of frequency lines of SO_2 concentrations in a logarithmic probability diagram

appendix 34

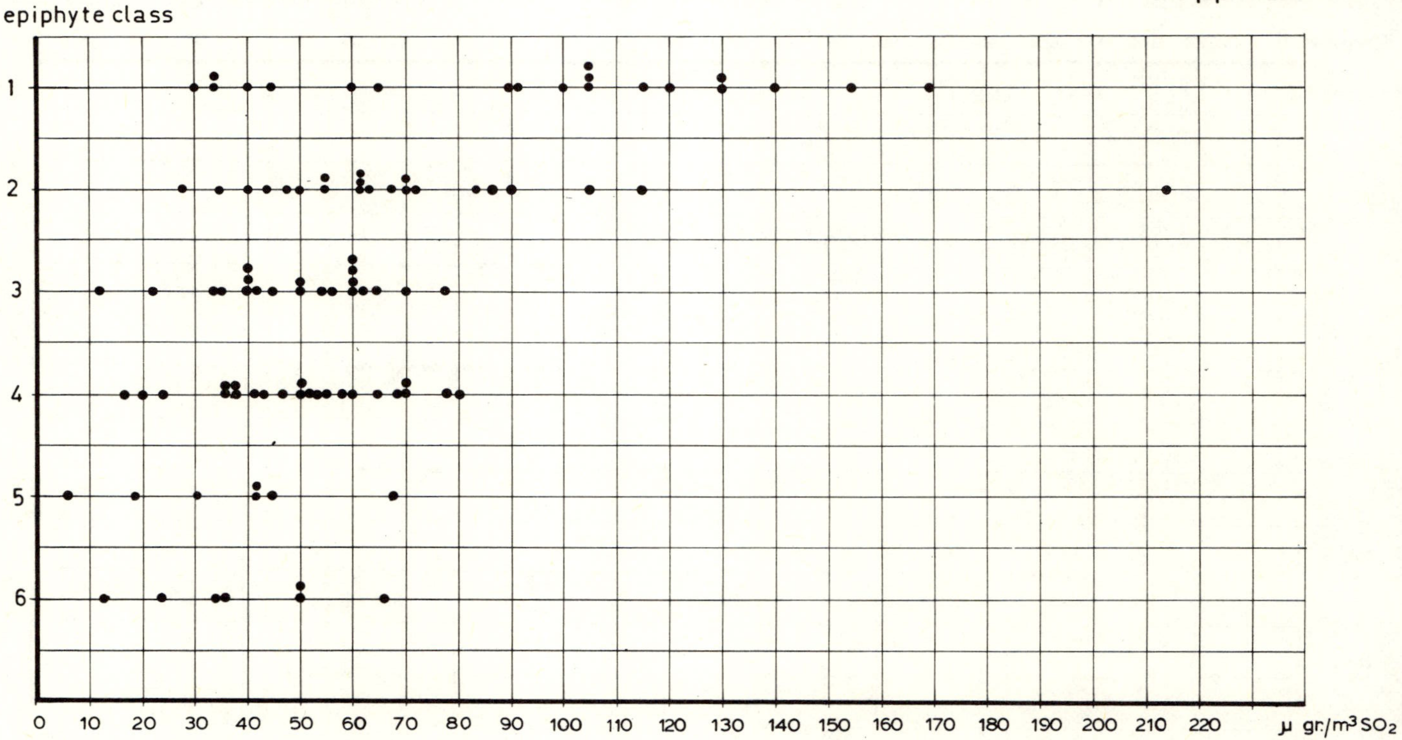

correlation between epiphyte class and 50 percentile values of SO_2 concentrations for 100 squares

appendix 35

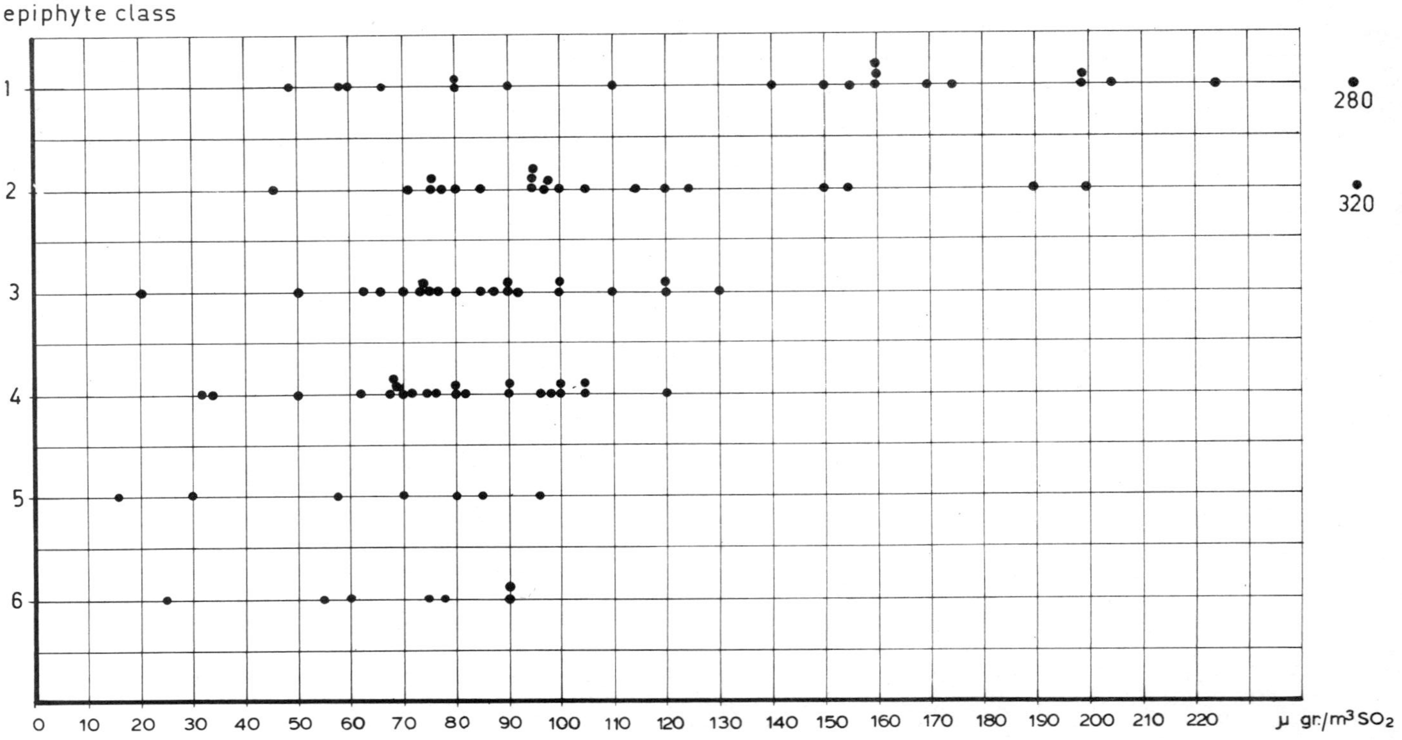

correlation between epiphyte class and 75 percentile values of SO_2 concentrations for 100 squares

appendix 36

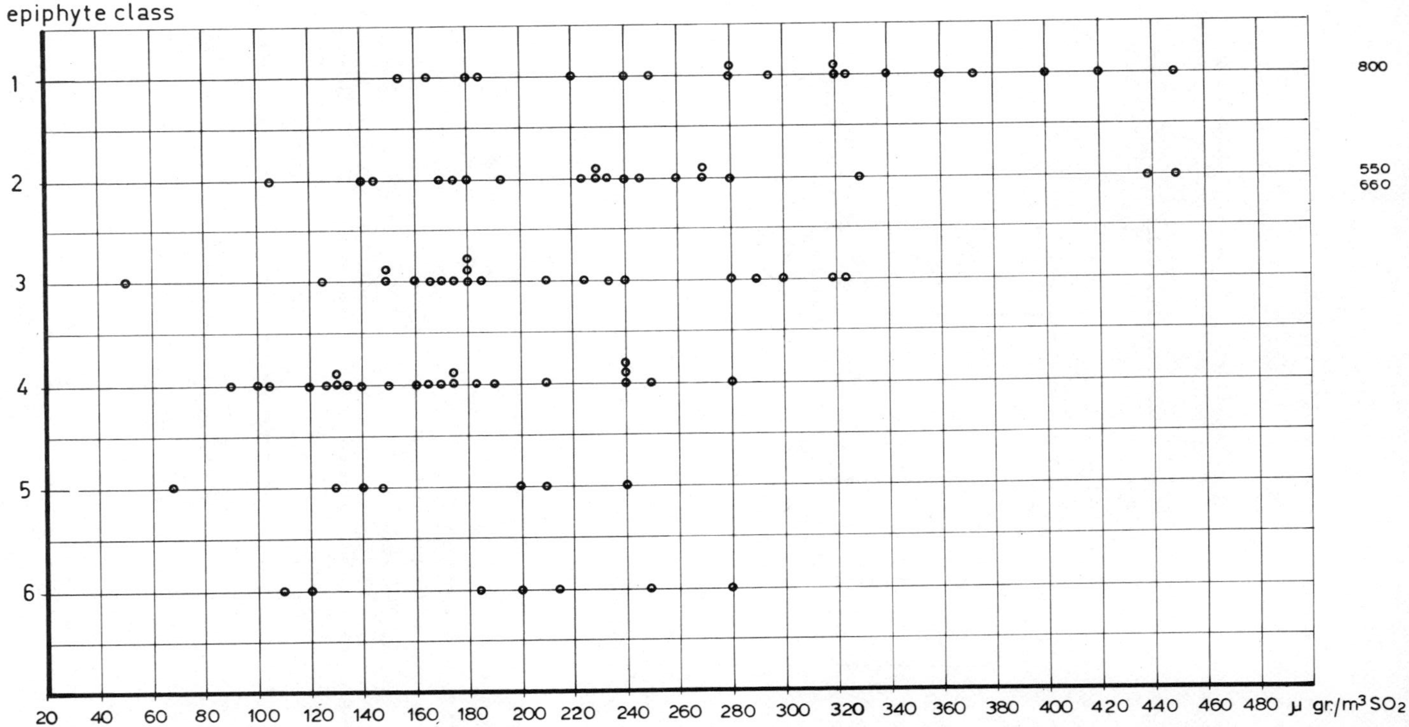

correlation between epiphyte class and 98 percentile values of SO_2 concentrations for 100 squares

appendix 37

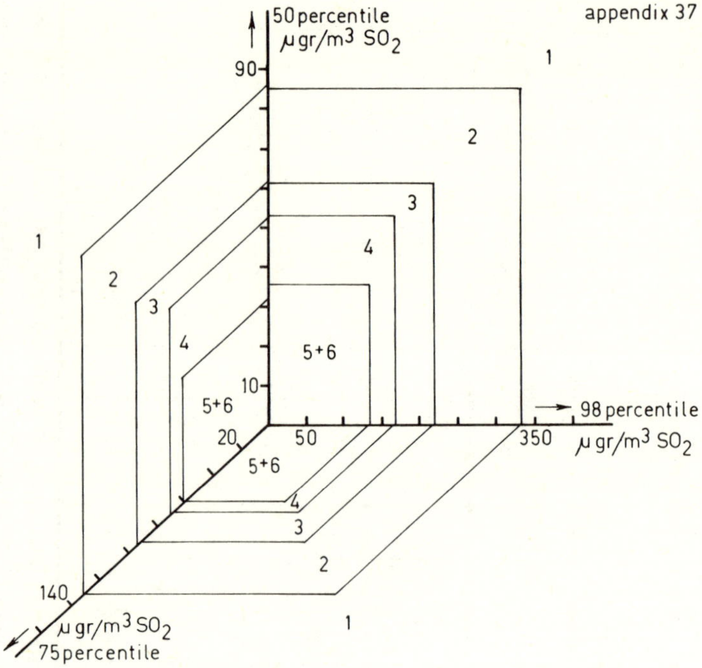

Correlation between epiphyte class and SO_2 concentration according to the combination I model.

Class limits are the same as above except for the 50 percentile: in this model 35 μg instead of 32 μg.

1-6: predicted epiphyte classes.

Class limits of SO_2 concentrations used to predict epiphyte classes.

Concentrations given in μg/m^3 SO_2.

predicted class	50 percentile	75 percentile	98 percentile
5 + 6	0 - 32	0 - 60	0 - 125
4	33 - 52	61 - 72	126 - 165
3	53 - 60	73 - 94	166 - 219
2	61 - 87	95 - 129	220 - 335
1	≽ 88	≽ 130	≽ 336

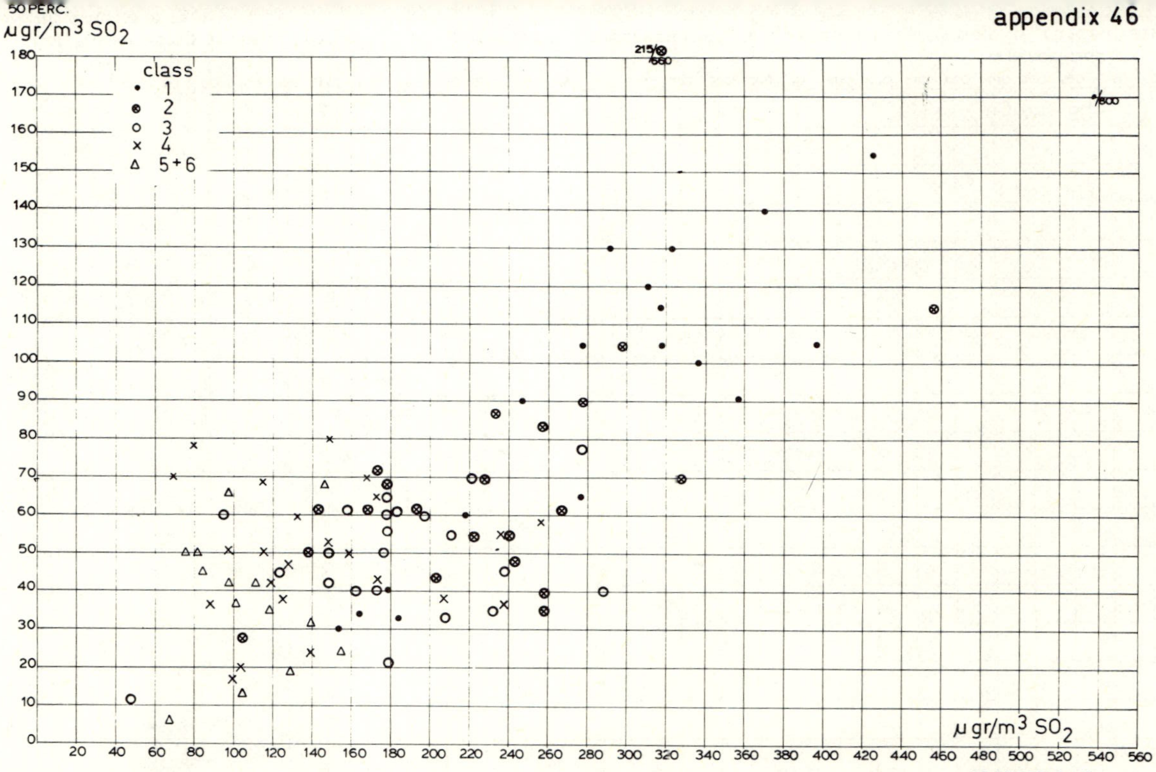

Correlation diagram of 50 percentile, 98 percentile, and observed epiphyte classes after correction of SO_2 concentrations for coastal effect.

appendix 47

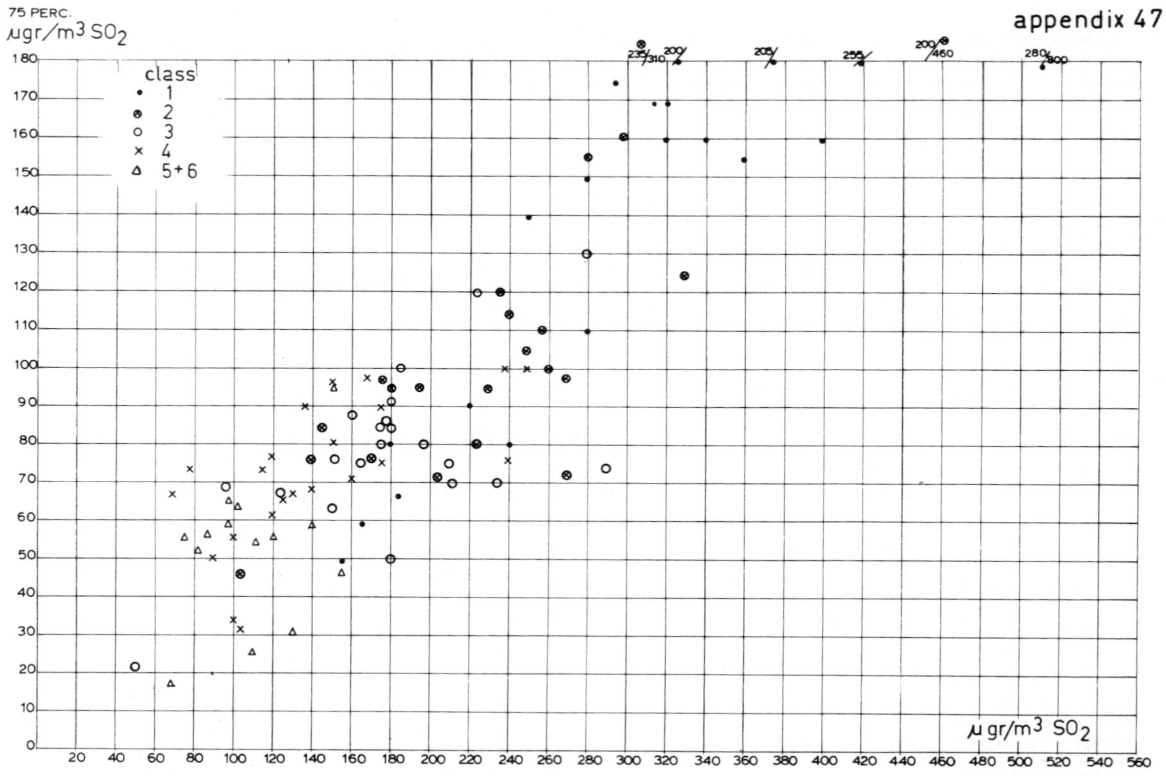

Correlation -diagram of 75 percentile, 98 percentile and **observed** epiphyte **classes** after correcting SO_2 concentrations for coastal effect

appendix 48

MAP SHOWING DISCREPANCIES BETWEEN PREDICTED (with combination 50/98 II) AND OBSERVED EPIPHYTE CLASSES AFTER CORRECTION FOR COASTAL EFFECT

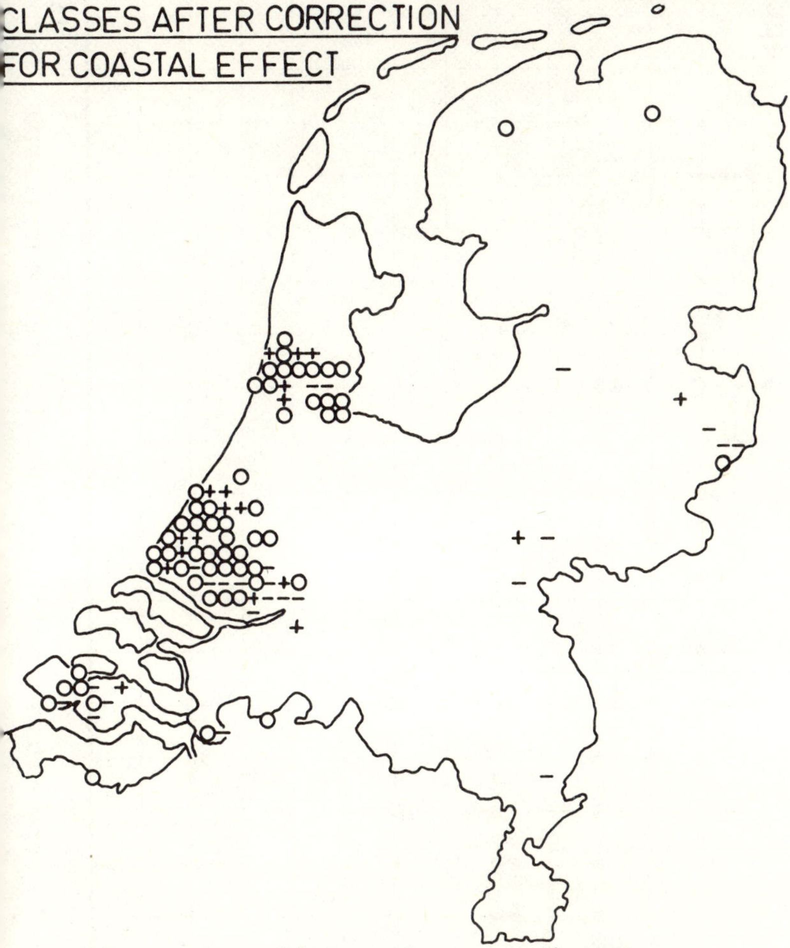

+ = epiphyte class higher than predicted
– = epiphyte class lower than predicted
o = epiphyte class same as predicted

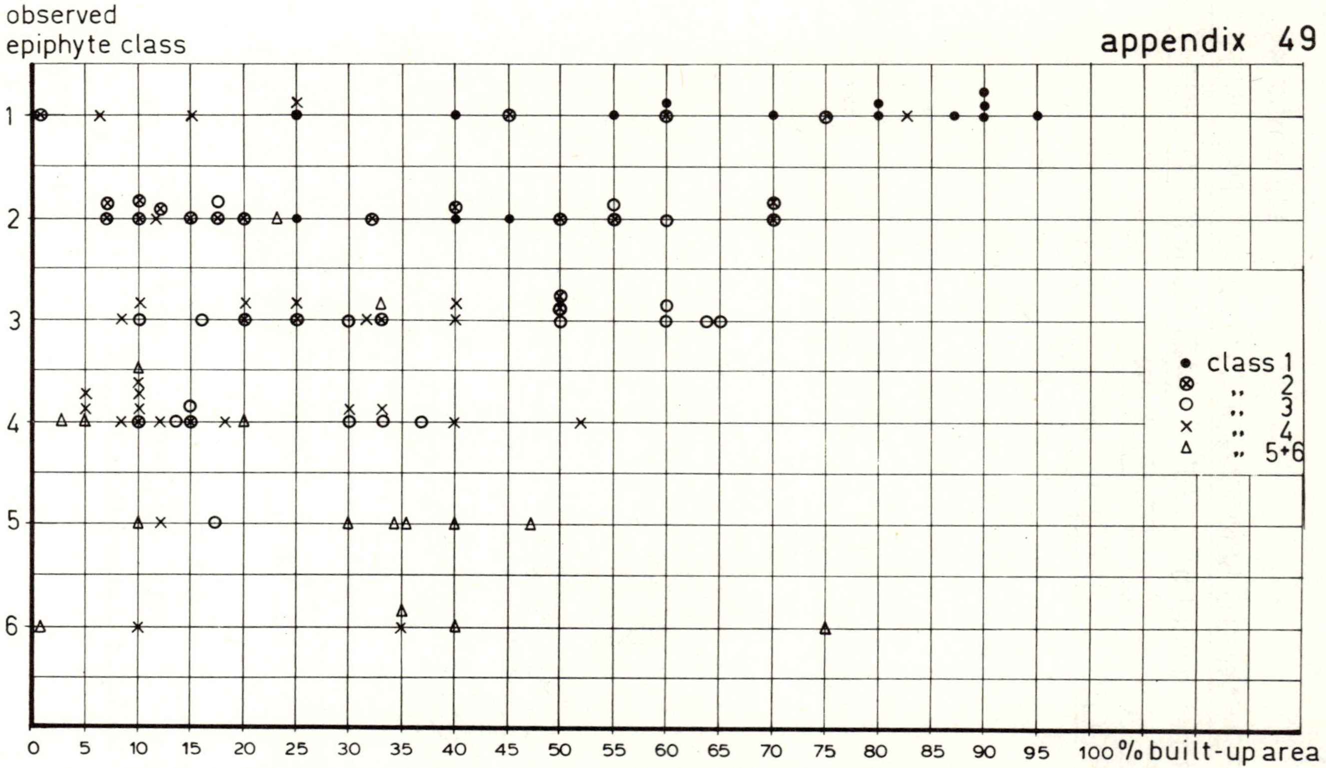

△ – ● : predicted epiphyte class (combination 50/98 II, corrected for coastal effect)

appendix 50

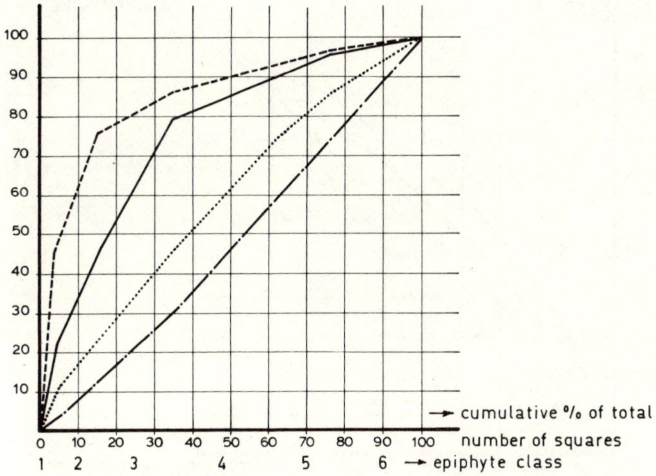

	number of inhabitant per square	% at 50 % of total number of squares	n
——— (solid)	> 25,000	86	65
········ (dotted)	10,000 - 25,000	61	131
—·—· (dash-dot)	< 10,000	46	1199
- - - - (dashed)	> 25,000 inhabitants and > 5,000 industrial employees	90	57

appendix 51

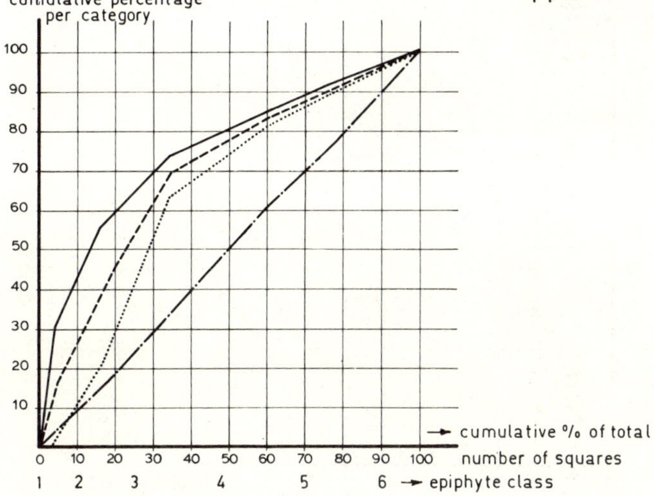

	number of industrial employees in the municipality	% at 50 % of total number of squares	n
——— (solid)	> 5,000	80	46
- - - - (dashed)	> 3,000	78	87
········ (dotted)	3,000 - 5,000	74	41
—·—· (dash-dot)	< 3,000	50	1308

appendix 52

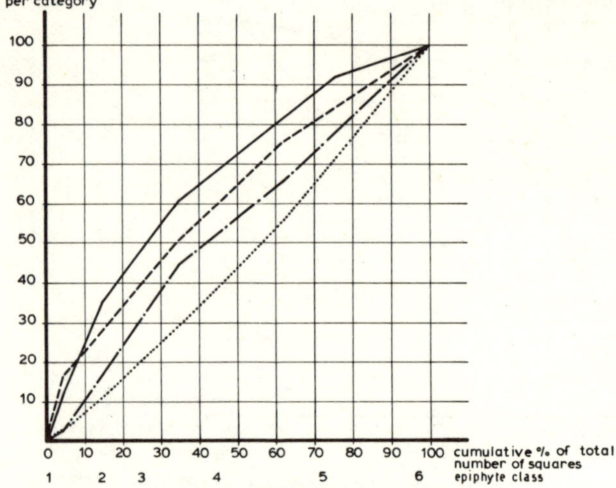

	number of factories	number of inhabitants	% at 50 % of total number of squares	n
················	< 2	10,000	44	1,047
—·—·—	> 2	10,000	57	150
- - - - - -	< 2	10,000	65	85
————	> 2	10,000	73	113

appendix 53

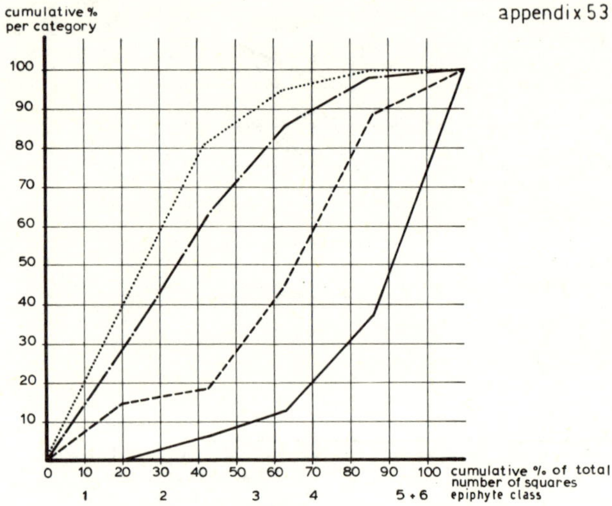

	predicted epiphyte class (combination 50/98 II, corrected for coastal effect)	% at 50 % of total number of squares	n
················	1 t/m 2	88	42
—·—·—	1 t/m 3	72	63
- - - - - -	4	38	23
————	5 en 6	9	14

appendix 54

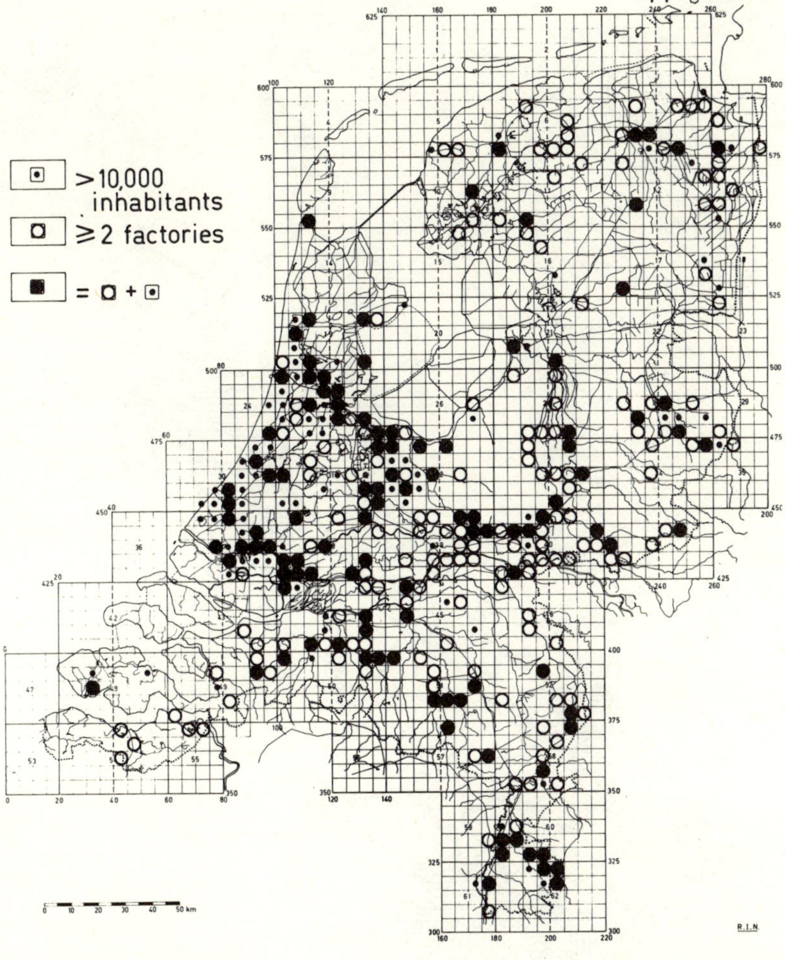

appendix 55

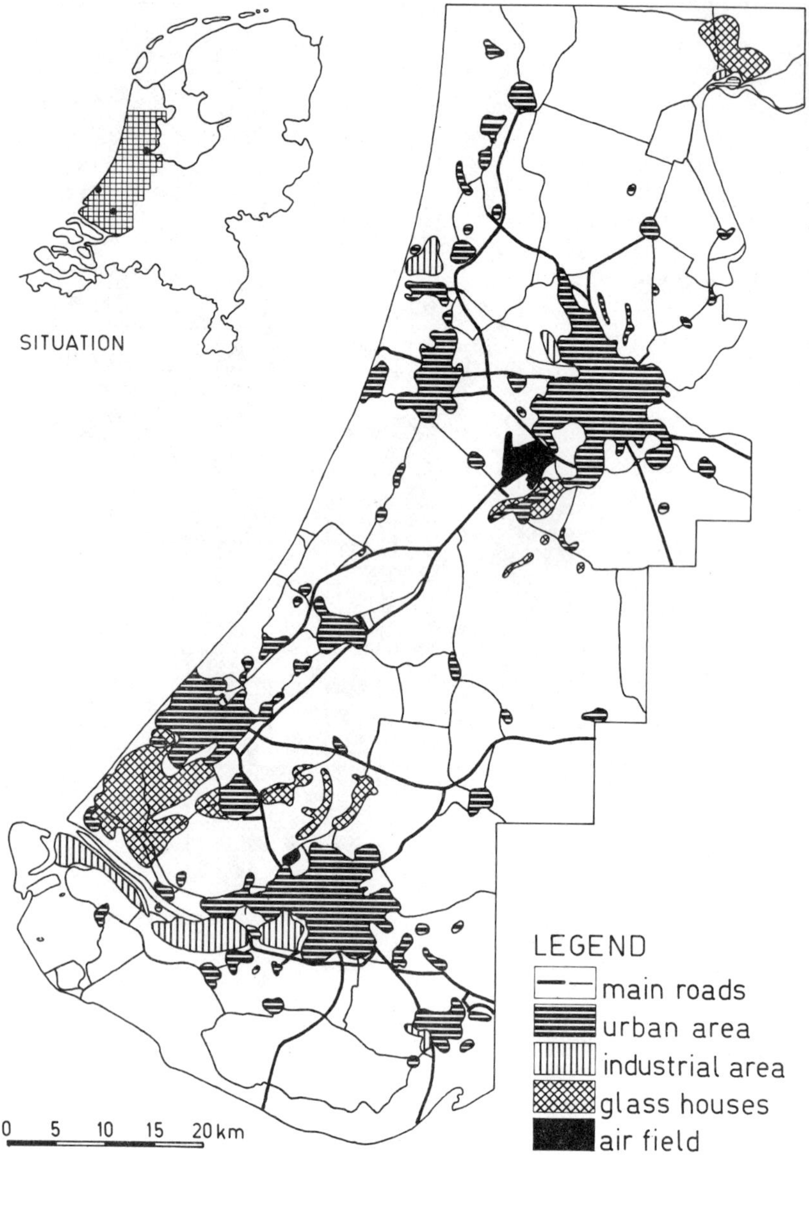

SITUATION
LEGEND
main roads
urban area
industrial area
glass houses
air field
0 5 10 15 20 km

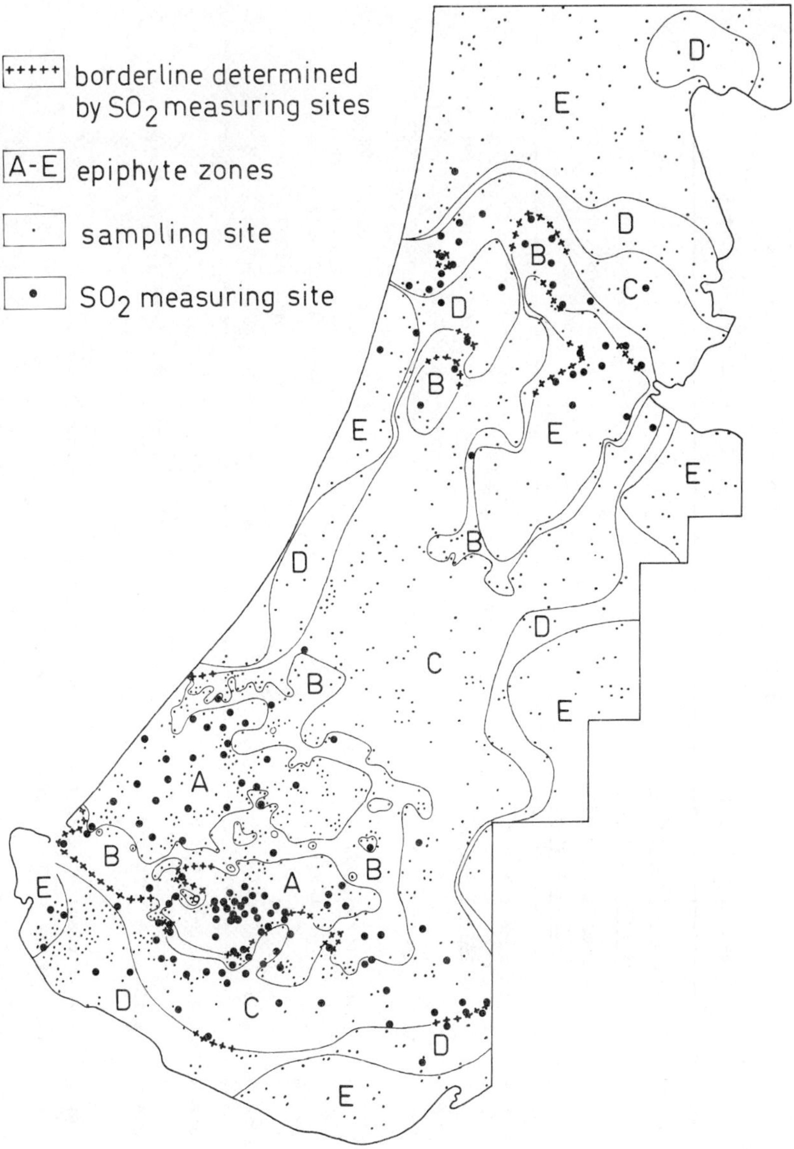
+++++ borderline determined by SO_2 measuring sites
A-E epiphyte zones
sampling site
SO_2 measuring site

Zone map of epiphytic lichens in The Netherlands

	sampling site with				SO_2 measuring site with	
Zone A	only: <u>Lecanora conizaeoides</u>, <u>Buellia punctata</u>, <u>Lepraria incana</u>, <u>Lecanora dispersa</u>			●	▲	in at least 1 out of 2 (3) winters 50 percentile values ⩾ 100 ug/m^3 SO_2
Zone B	as in zone A + <u>Lecanora expallens</u>			◑	▲	in at least 1 out of 2 (3) winters 75 percentile values ⩾ 100 ug/m^3 SO_2
	number of lichen species on:					
	Ulmus	Populus, Fraxinus, Tilia	Salix			
Zone C	0-11	0-6	0-7	⊙	◬	in at least 1 out of 2 (3) winters 98 percentile values ⩾ 150 ug/m^3 SO_2
Zone D	12,13,14	7,8,9	8,9	○	△	less SO_2 than ▲ , ▲ or ◬
Zone E	⩾ 15	⩾ 10	⩾ 10	◠		

A site is always classified in the poorest (most polluted) group to which it can belong on the basis of the above definitions.

appendix 59

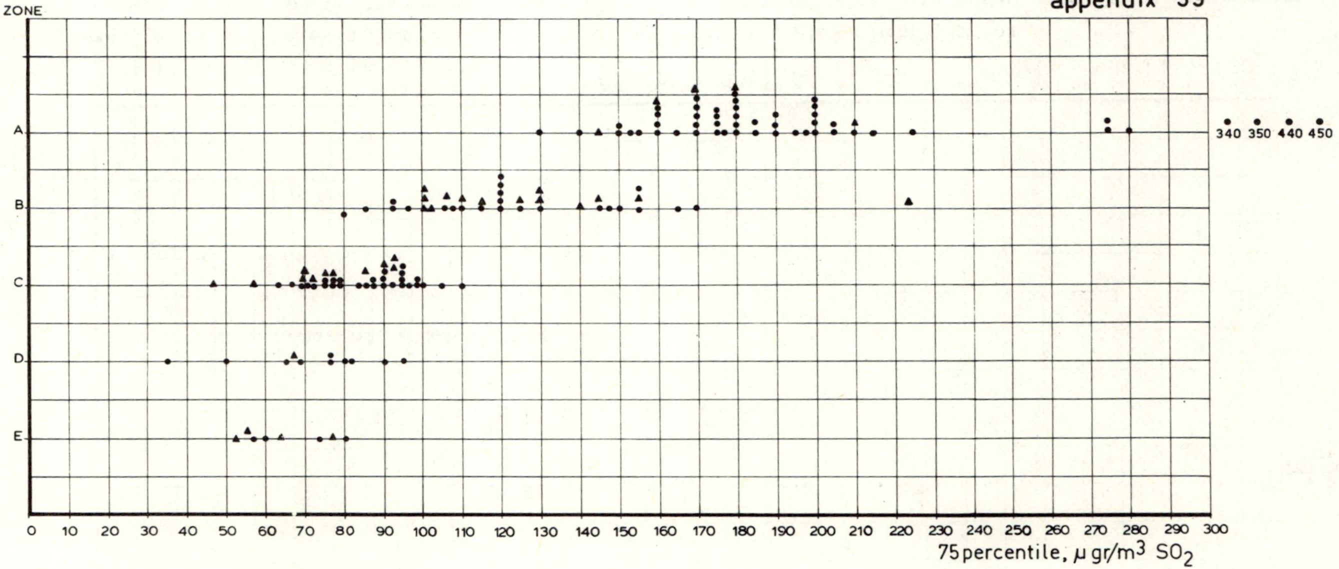

ZONE
A
B
C
D
E
0 10 20 30 40 50 60 70 80 90 100 110 120 130 140 150 160 170 180 190 200 210 220 230 240 250 260 270 280 290 300
340 350 440 450
75 percentile, μ gr/m3 SO2

appendix 60

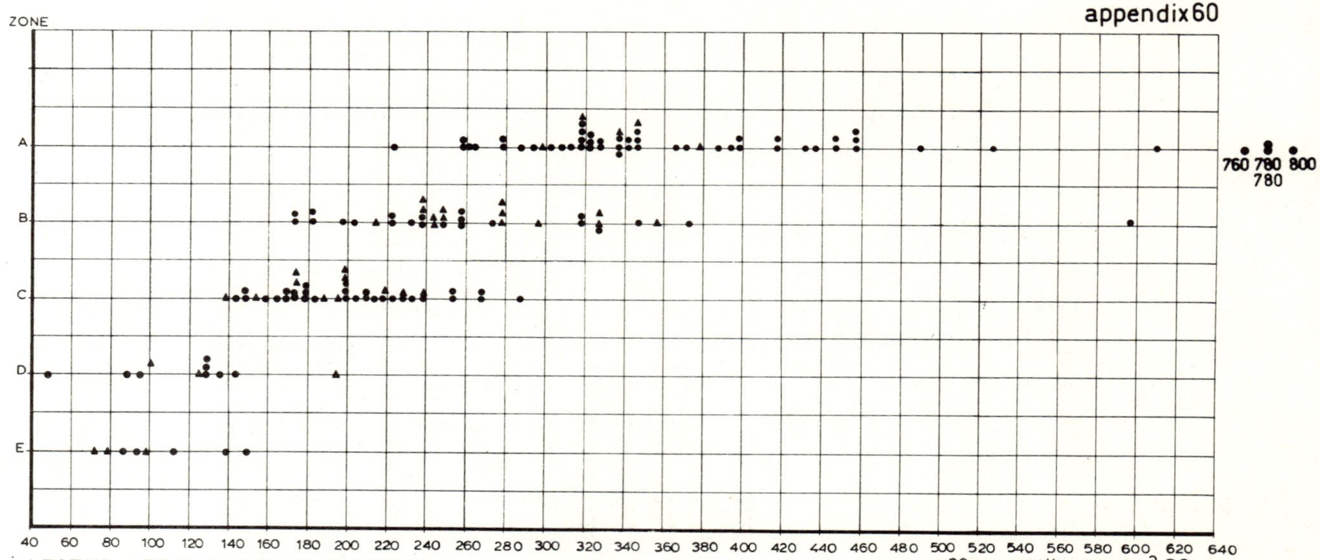

LEGEND APPENDICES 58, 59 AND 60 :

● = SO_2 measuring sites in Zuid-Holland ▲ = SO_2 measuring sites in Noord-Holland

For each measuring site the epiphyte zone in wich it is found is indicated Zone borders were drawn solely on the basis of lichen sampling sites, except for zone borders A/B and B/C in Noord-Holland The concentrations are the highest values per percentile for each SO_2 measuring site.

Comparison of the correlation of different percentiles with lichen zones
For explanation see text, section 10.3 and 10.5

61 a

percentile	50	75	98
zone border			
A/B	100	145	285
B/C	70	100	240
C/D	50	70	150
D/E	50	62	120

SO_2-concentrations (in $\mu g/m^3$) corresponding with lichen zone borders; concentrations are chosen such that overlap (see 61 b) is minimal.

61 b

Numbers of "overlapping" SO_2 sites, using limiting concentrations given in 61 a

percentile	50	75	98
zone border			
Z-Holland A/B	[6]	8	13
Z-Holland B/C	10	[7]	17
Z-Holland C/D	14	9	[2]
Z-Holland + N-Holland C/D	25	13	[4]
Z-Holland + N-Holland D/E	26	8	[6]

total number of sites per zone:

Z-Holland	zone A	51
Z-Holland	zone B	20
Z-Holland	zone C	30
Z+N-Holland	zone C	44
Z+N-Holland	zone D	11
Z+N-Holland	zone E	8

borders of lichen zones are determined on the basis of lichen sampling sites only (this was not possible for borders A/B and B/C in Noord-Holland).

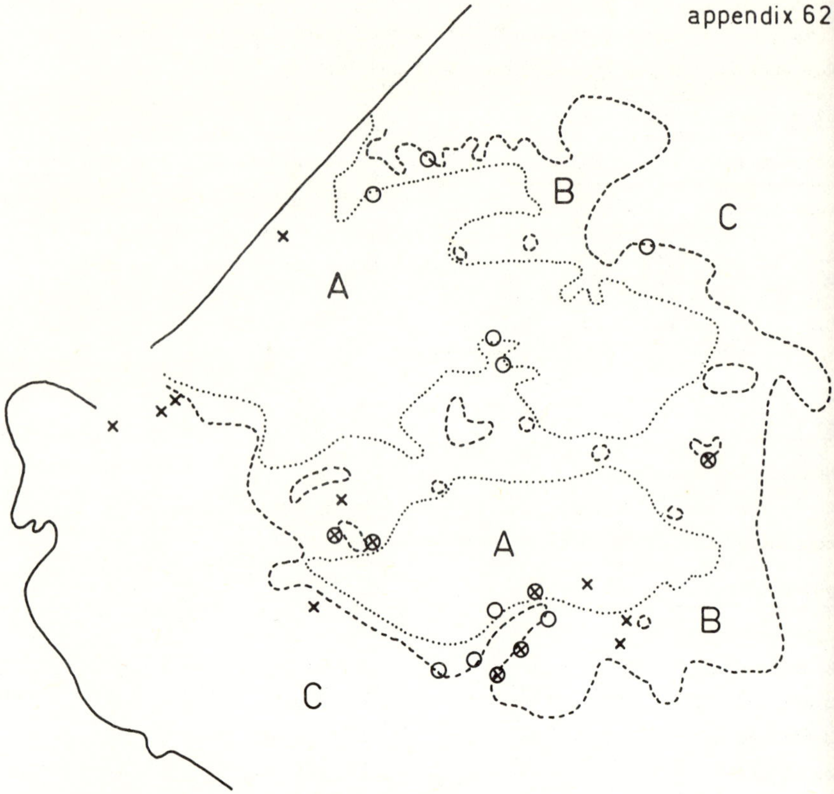

× measuring site with a SO_2 concentration that doesn't suit with the lichen zone

○ measuring site on zone-border, SO_2 concentration suits with poorest zone

⊗ measuring site on zone-border, SO_2 concentration suits with richest zone

······ zone border A/B } The drawing of the zone borders has been done
---- zone border B/C } taking into account for the lichen distribution only

Comparison of the correlation of SO_2 concentrations of different seasons with lichen zones.
For explanation see text, section 10.3

zone border:	SO_2 values of:	percentile used for limiting concentration 50-	75-	98-
	winter	6	7	9
A/B	year	9	9	9
	summer	14	10	11
	winter	10	5	11
B/C	year	11	10	11
	summer	18	16	13
	winter	14	7	2
C/D	year	18	11	7
	summer	-	14	12
	winter	17	3	4
D/E	year	-	-	-
	summer	-	-	-

Number of SO_2 sites in Zuid-Holland with a SO_2 concentration that is not in accordance with the lichen zone. Compare Appendix 61 b. Numbers of overlapping sites are lower here, because sites on which coastal effect is applicable are disregarded. (Nothing is known about the coastal effect on summer or year values). Where overlapping was so large that determining a limiting concentration was senseless, no number is given.
Zone borders are as in Appendix 61 b (i.e. determined on the basis of lichen-sites only).
Number of SO_2-sites without coastal effect in Z-Holland: Zone A: 45, B: 15, C: 24, D: 8, E: 2.

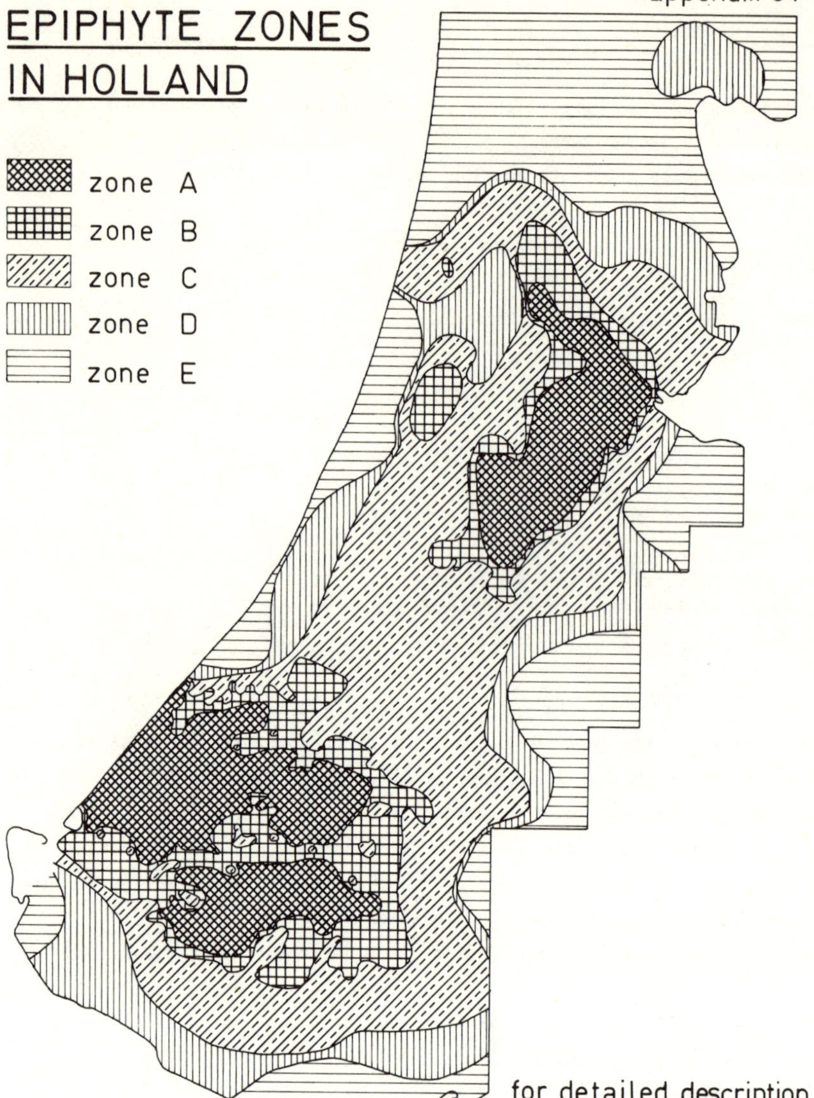
EPIPHYTE ZONES
IN HOLLAND
zone A
zone B
zone C
zone D
zone E
for detailed description
of zones, see appendix 57

LECANORA EXPALLENS

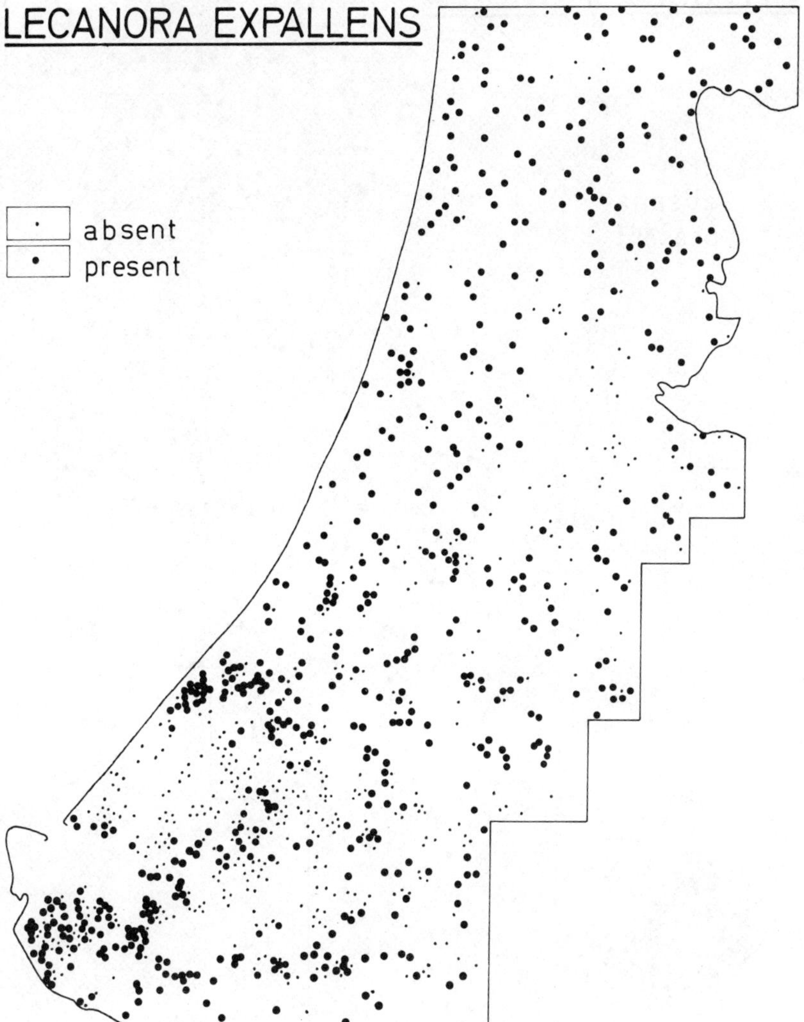

PHYSCIA TENELLA

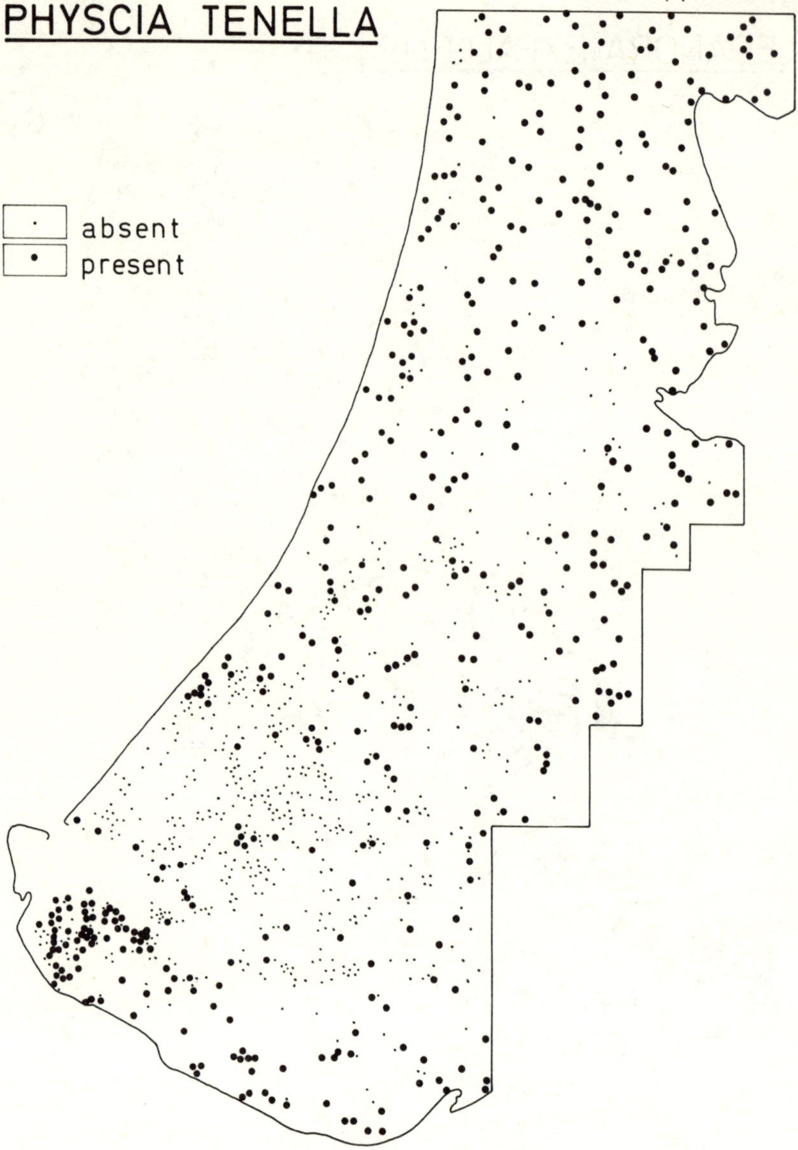

PARMELIA SULCATA

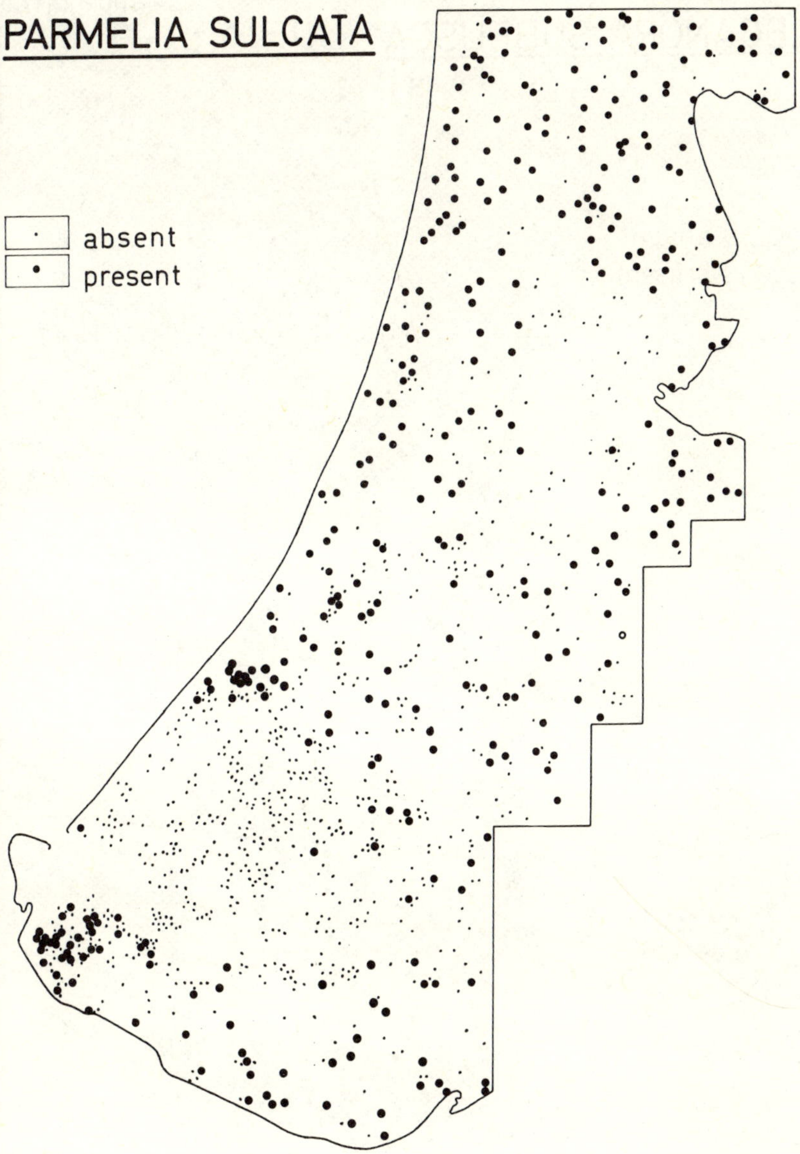

LECANORA SUBFUSCA s.l.

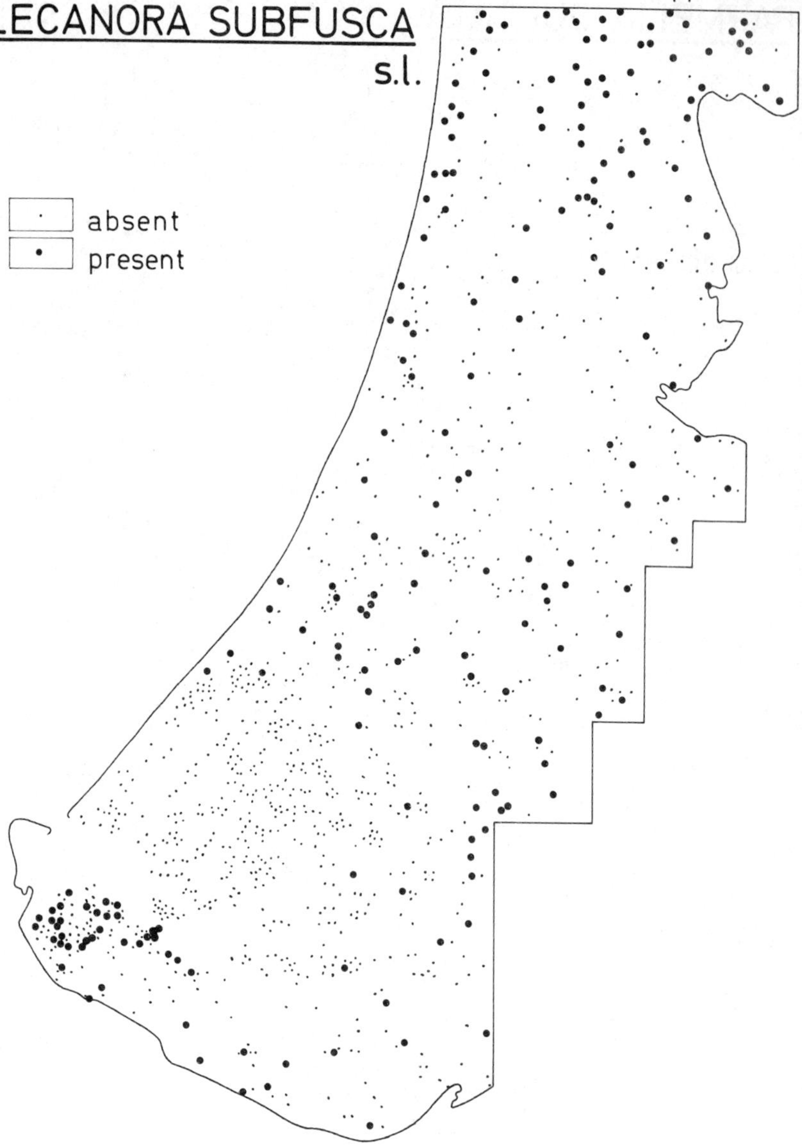

RAMALINA FARINACEA

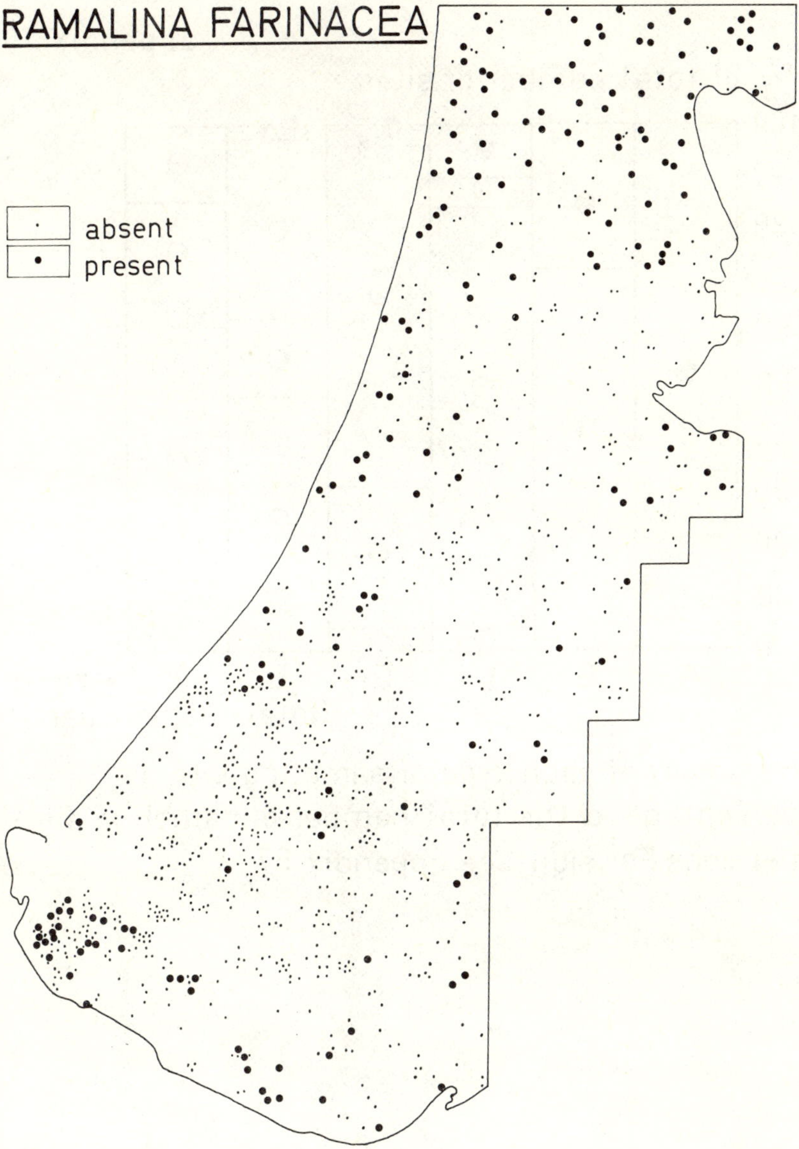

appendix 70

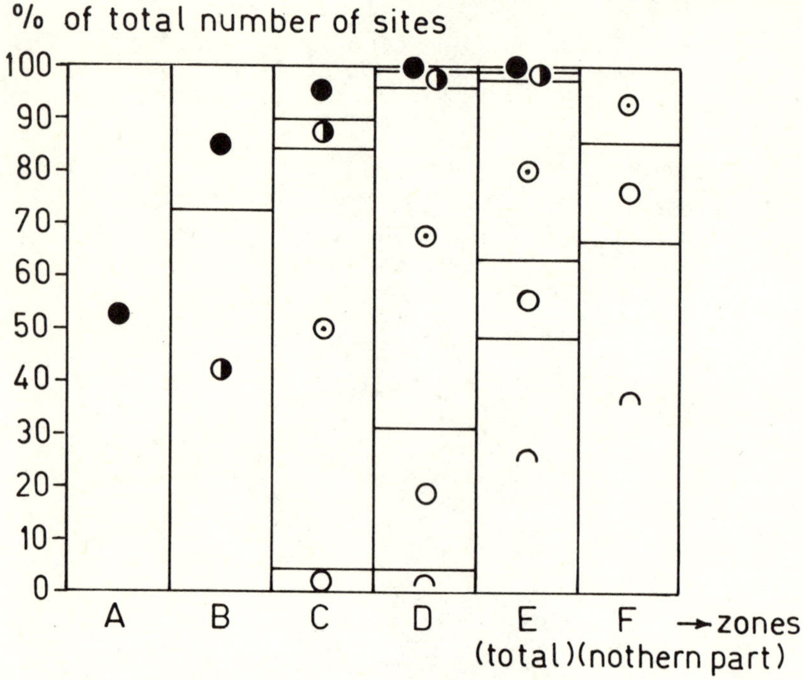

frequency of each type of sampling site as percentage of the total number of sampling sites per zone. For sign, see appendix 57.

appendix 71

concentrations are the highest winter values corrected for the coastal effect. Zone-borders as in appendix 57

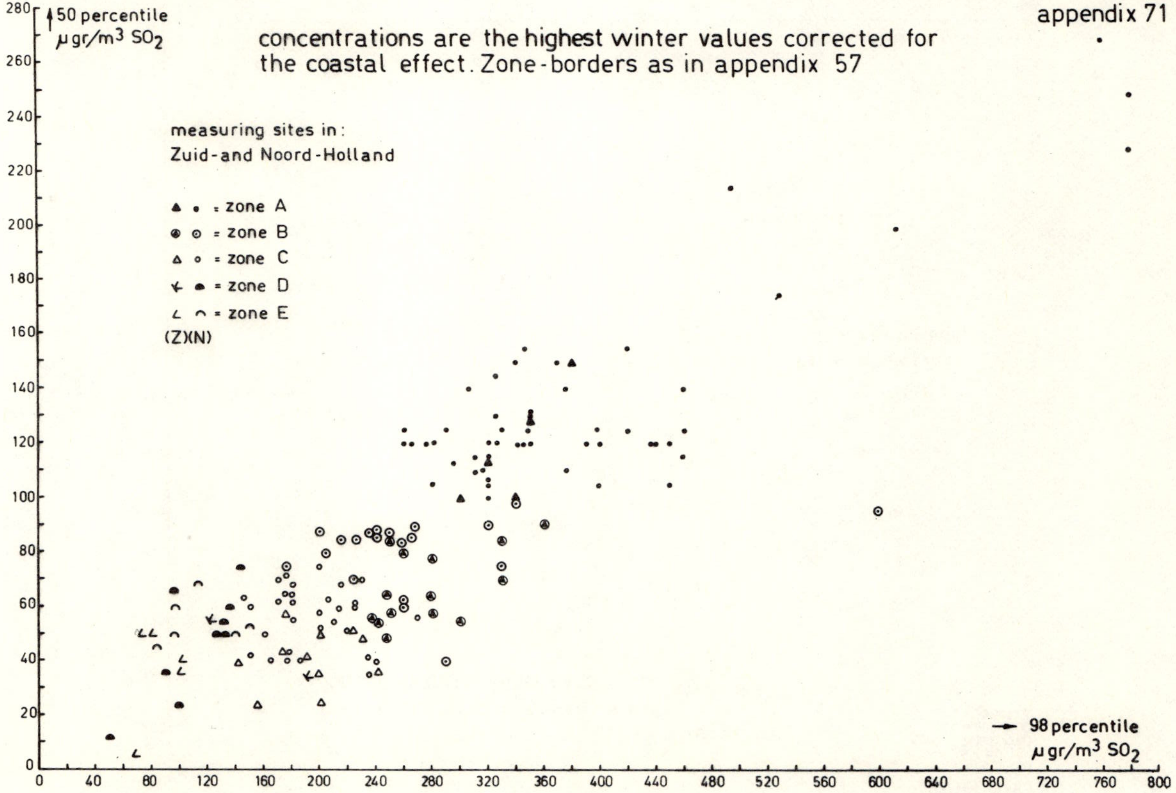

appendix 72

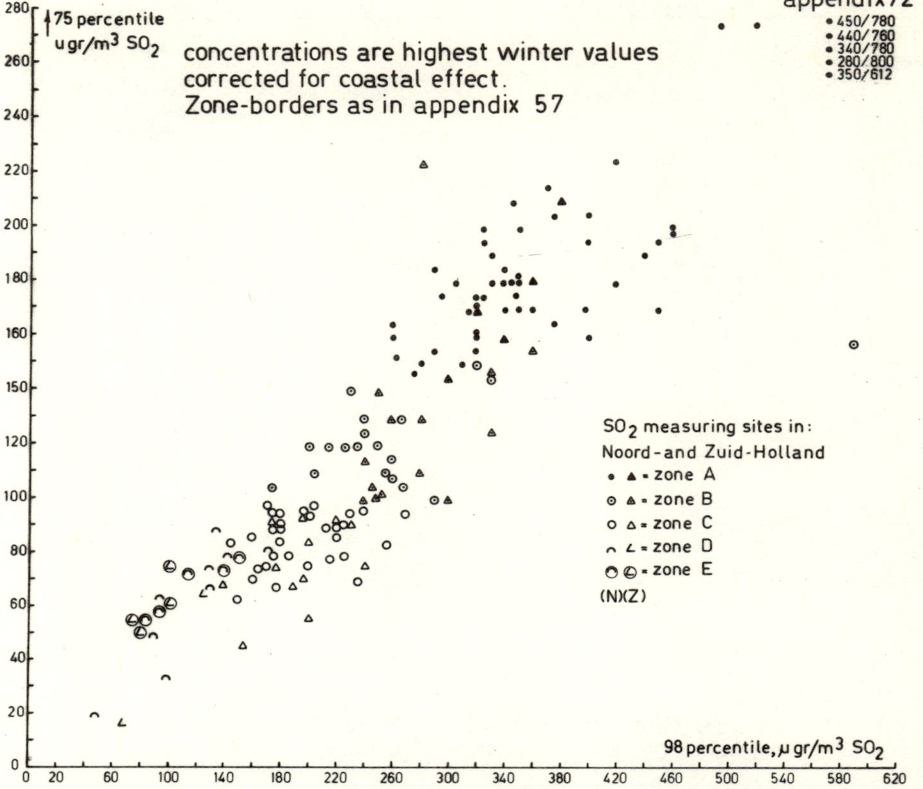

75 percentile ugr/m3 SO2
concentrations are highest winter values
corrected for coastal effect.
Zone-borders as in appendix 57
450/780
440/760
340/780
280/800
350/612
SO2 measuring sites in:
Noord-and Zuid-Holland
zone A
zone B
zone C
zone D
zone E
(N)(Z)
98 percentile, µgr/m3 SO2
0
20
40
60
80
100
120
150
160
180
200
220
240
260
280
0
20
60
100
140
180
220
260
300
340
380
420
460
500
540
580
620

MODELS FOR PREDICTING EPIPHYTE ZONES AND CLASSES FROM SO_2 CONCENTRATIONS

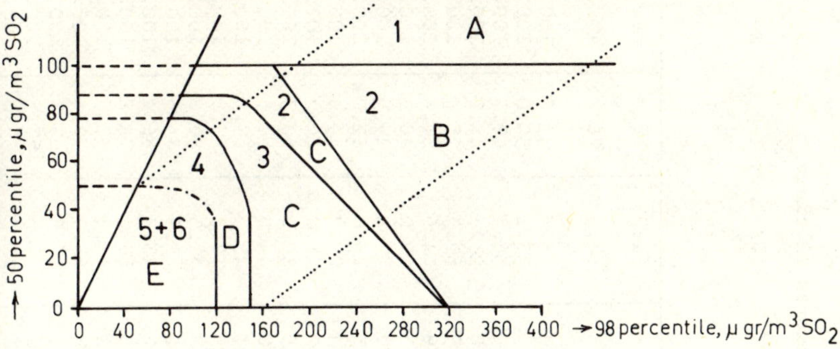

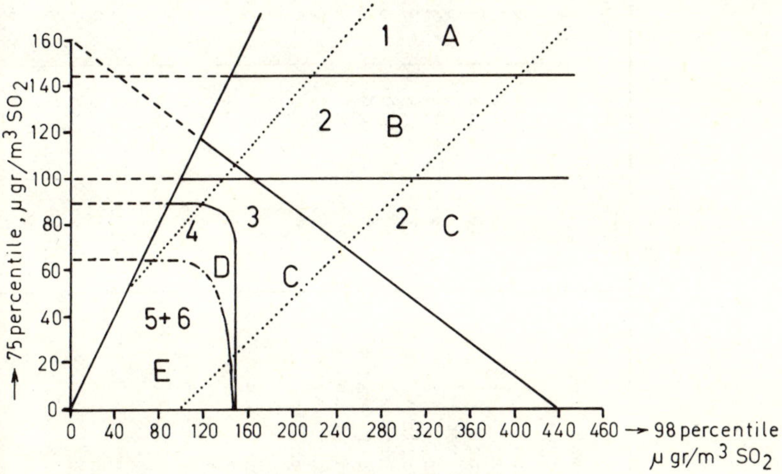

difference predicted/ observed	model 50/98 III	75/98 III
0 classes	57	58
1 class	32	30
	89	88
2 classes	10	11
3 classes	1	1
4 classes	0	0
chi^2	120,2	114,9

see appendix 39

----- values impossible (because 75 or 50 percentile is always < 98 percentile)

..... limits of observation scatter

—·—·—uncertain

	total number of ***) squares	percentage squares in class: 1	1-2	1-3	1-4	1-5	1-6
All squares	1,395	4.1	15.4	34.6	61.8	76.0	100.0
All squares except those incompletely surveyed	1,310	4.1	16.0	34.1	61.9	76.0	100.0
Arthonia radiata	30	0.0	0.0	13.3	33.3	43.3	100.0
Arthonia spadicea	28	0.0	3.6	14.3	32.1	50.0	100.0
Buellia canescens	468	1.1	7.9	24.6	56.2	69.7	100.0
Buellia griseovirens	229	0.0	0.0	4.8	21.8	44.1	100.0
Buellia punctata	1,339	2.8	14.1	32.9	60.9	75.4	100.0
Catillaria griffithii	137	1.5	6.6	13.9	44.5	61.3	100.0
Candelaria concolor *)	168	0.0	0.6	6.5	28.0	48.2	100.0
Candelariella "reflexa" *)	289	0.0	0.4	6.9	29.4	49.5	100.0
Candelariella vitellina *)	589	0.3	3.7	16.5	43.6	61.0	100.0
Candelariella xanthostigma *)	159	0.0	0.0	4.4	25.2	43.4	100.0
Cetraria chlorophylla	65	0.0	0.0	9.2	26.2	46.2	100.0
Cetraria glauca	63	0.0	11.1	36.5	46.0	60.3	100.0
Chaenotheca ferruginea	75	1.3	5.3	20.0	44.0	57.3	100.0
Evernia prunastri	931	0.1	5.7	21.7	49.3	67.5	100.0
Haematomma coccineum	120	0.0	1.7	7.5	17.5	39.2	100.0
Lecanora carpinea	423	0.0	1.4	8.3	31.5	50.8	100.0
Lecanora conizaeoides	1,243	4.1	15.9	35.2	61.6	75.6	100.0
Lecanora dispersa	339	2.1	6.8	21.2	51.3	66.4	100.0
Lecanora expallens	1,260	1.8	11.8	31.0	59.2	74.2	100.0
Lecanora laevis	72	0.0	0.0	4,2	13.9	37.5	100.0
Lecanora muralis	97	0.0	9.3	25.8	68.0	79.4	100.0
Lecanora subfusca s.l.	1,069	0.7	7.2	22.9	52.9	69.7	100.0
Lecidea granulosa	40	0.0	2.5	20.0	37.5	57.5	100.0
Lecidea limitata	541	0.7	3.7	14.1	41.2	59.0	100.0

Lecidea uliginosa	45	0.0	4.7	14.0	50.2	55.5	100.0
Lepraria incana	1,230	2.4	12.8	32.0	59.4	74.3	100.0
Parmelia acetabulum	733	0.0	2.3	12.6	41.5	60.8	100.0
Parmelia caperata	55	0.0	0.0	3.6	16.4	27.3	100.0
Parmelia exasperatula	326	0.0	0.3	5.5	27.7	47.9	100.0
Parmelia furfuracea	173	0.0	1.7	10.4	34.7	54.7	100.0
Parmelia laciniatula	42	0.0	2.4	2.4	9.5	23.8	100.0
Parmelia physodes	890	0.6	8.4	24.4	51.7	69.0	100.0
Parmelia revoluta	55	0.0	0.0	3.6	9.1	29.1	100.0
Parmelia saxatilis	84	0.0	2.4	17.9	25.0	45.3	100.0
Parmelia subaurifera **)	586	0.3	2.1	12.0	36.9	56.3	100.0
Parmelia subrudecta	683	0.4	4.6	14.1	40.4	59.6	100.0
Parmelia sulcata	1,219	0.8	8.8	27.3	56.8	72.8	100.0
Parmelia tiliacea	51	2.0	3.9	5.9	27.5	45.1	100.0
Parmelia tubulosa	84	0.0	1.2	15.5	25.0	45.2	100.0
Pertusaria albescens	213	0.0	1.9	7.5	21.6	40.4	100.0
Pertusaria amara	187	0.0	2.7	10.7	28.9	44.9	100.0
Pertusaria coccodes	58	0.0	0.0	3.5	13.8	32.8	100.0
Pertusaria pertusa	108	0.0	0.9	7.4	17.6	37.0	100.0
Phlyctis argena	297	0.0	0.4	4.0	18.9	38.0	100.0
Physcia adscendens	247	0.0	4.1	18.2	41.3	59.1	100.0
Physcia caesia	468	0.4	2.4	13.7	42.3	61.1	100.0
Physcia grisea	551	0.5	3.4	17.1	43.9	60.2	100.0
Physcia orbicularis	360	0.6	3.6	15.8	42.2	62.8	100.0
Physcia pulverulenta	70	0.0	2.9	10.0	22.9	38.6	100.0
Physcia tenella (incl. adsc.)	1.269	1.7	11.1	29.9	58.9	74.1	100.0
Ramalina farinacea	360	0.6	3.6	15.8	42.2	62.8	100.0
Ramalina fastigiata	408	0.0	0.7	5.6	24.8	46.1	100.0
Ramalina fraxinea	98	0.0	1.0	2.0	10.2	32.7	100.0
Usnea spp.	76	0.0	0.0	10.5	28.9	46.0	100.0
Xanthoria candelaria	862	0.6	4.4	18.8	48.0	65.9	100.0
Xanthoria parietina	714	0.4	5.0	19.3	46.8	64.9	100.0
Xanthoria polycarpa	818	0.5	3.6	20.5	49.2	67.9	100.0

*) According to Brand (personal communication) most of _Candelaria concolor_ records probably belong to _Candelariella reflexa_ and most of _Candelariella xanthostigma_ records are either _Candelariella reflexa_ or _Candelariella vitellina_. Here the indication _Candelariella "reflexa"_ indicates all _Candelaria concolor_ and all _Candelariella xanthostigma_ recordings combined. _Candelariella vitellina_ records also include sterile _C.aurella_ specimens.

**) Because some of the _Parmelia glabratula_ records probably belong to _P. subaurifera_, these species are given together under the name _P.subaurifera_ (the majority actually being _P.subaurifera_)

***) The total number of squares given here is in most cases lower than the number of squares on the distribution maps, because these maps include also records from non-standard sites are included, which were excluded for the calculation of the sensitivity curves.

Nomenclature according to James 1965

appendix 75

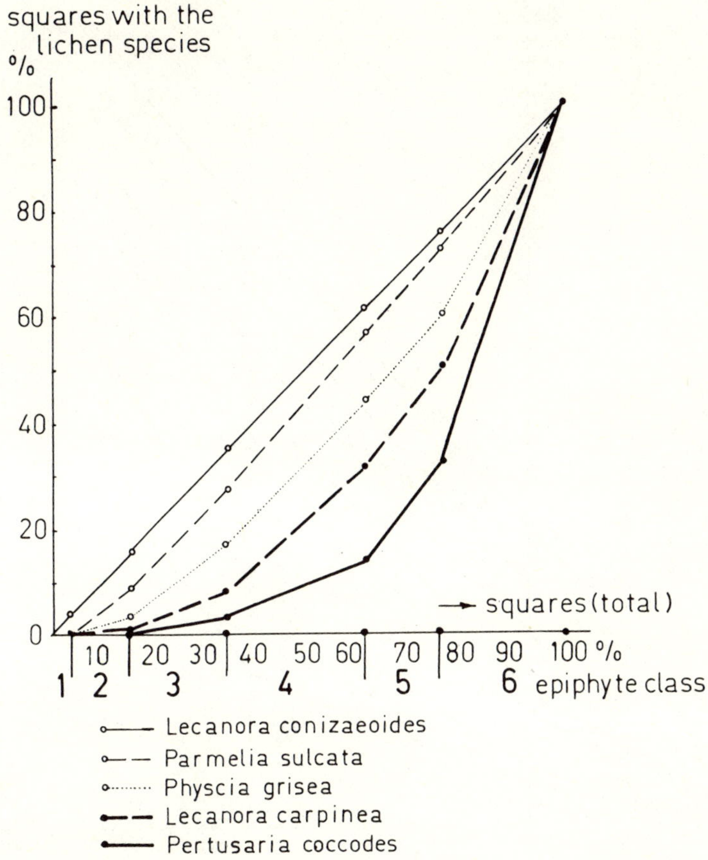

SENSITIVITY CURVES. Cumulative frequency percentages of the squares in which the lichen species has been found (vertical axis) plotted against cumulative frequency percentages of all the squares (per class).

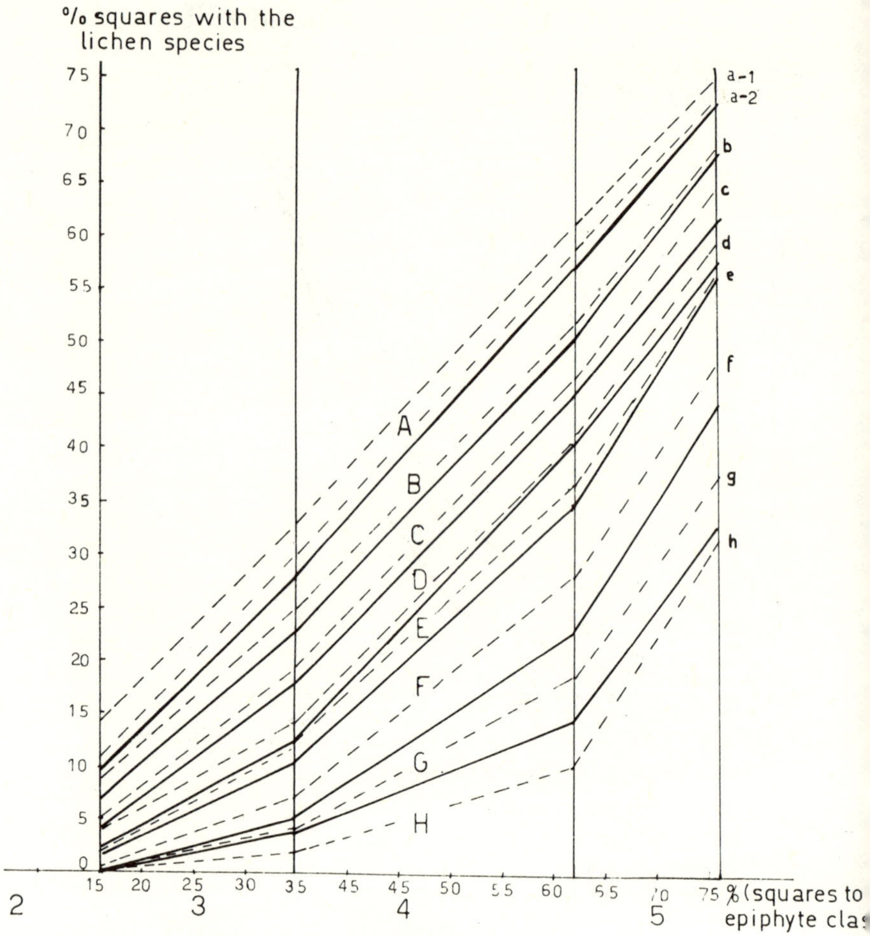

sensitivity class	sensitivity curve of	
A	Lecanora conizaeoides	(a 1)
A	Lecanora expallens	(a 2)
B	Physcia tenella	(b)
C	Parmelia physodes	(c)
D	Parmelia subrudecta	(d)
E	Parmelia subaurifera	(e)
F	Candelaria concolor	(f)
G	Phlyctis argena	(g)
H	Ramalina fraxinea	(h)

sensitivity class	species rankable without intersecting curves	species with sensitivity curves intersecting curves of other species, but not intersecting class limits	species with sensitivity curves intersecting class limits between (): species with a divergent curve, class uncertain.
	Lecanora conizaeoides		
	Buellia punctata		
A	Lepraria incana		
	Lecanora expallens		
	Physcia tenella		(Cetraria glauca)
	Parmelia sulcata		
	Buellia canescens		Lecidea scalaris
B	Parmelia physodes		
	Lecanora subfusca s.l.		
	Lecanora dispersa		
	Evernia prunastri	Xanthoria polycarpa	
C	Xanthoria candelaria		Lecanora muralis
	Xanthoria parietina		
	Physcia grisea		Physcia adscendens
	Candelariella vitellina	Physcia orbicularis	(Lecidea granulosa)
D	Ramalina farinacea	Parmelia subrudecta Physcia caesia	Chaenotheca ferruginea Catillaria griffithii
	Lecidea limitata	Parmelia acetabulum	
E	Parmelia subaurifera		Lecidea symmicta
	Parmelia furfuracea		{(Arthonia spadicea) (Arthonia radiata)
			Pertusaria amara
	Lecanora carpinea	Usnea spp.	Lecidea uliginosa (Parmelia saxatilis)
		Parmelia tubulosa	Parmelia tiliacea
F	Candelaria concolor	Parmelia exasperatula	Haematomma coccineum
		Candelariella xanthostigma	Physcia pulverulenta
	Ramalina fastigiata		Pertusaria albescens
		Cetraria chlorophylla	
G	Buellia griseovirens		Pertusaria pertusa
	Phlyctis argena	Lecanora laevis	
	Pertusaria coccodes	Parmelia revoluta	
H	Ramalina fraxinea	Parmelia caperata	
		Parmelia laciniatula	

appendix 78

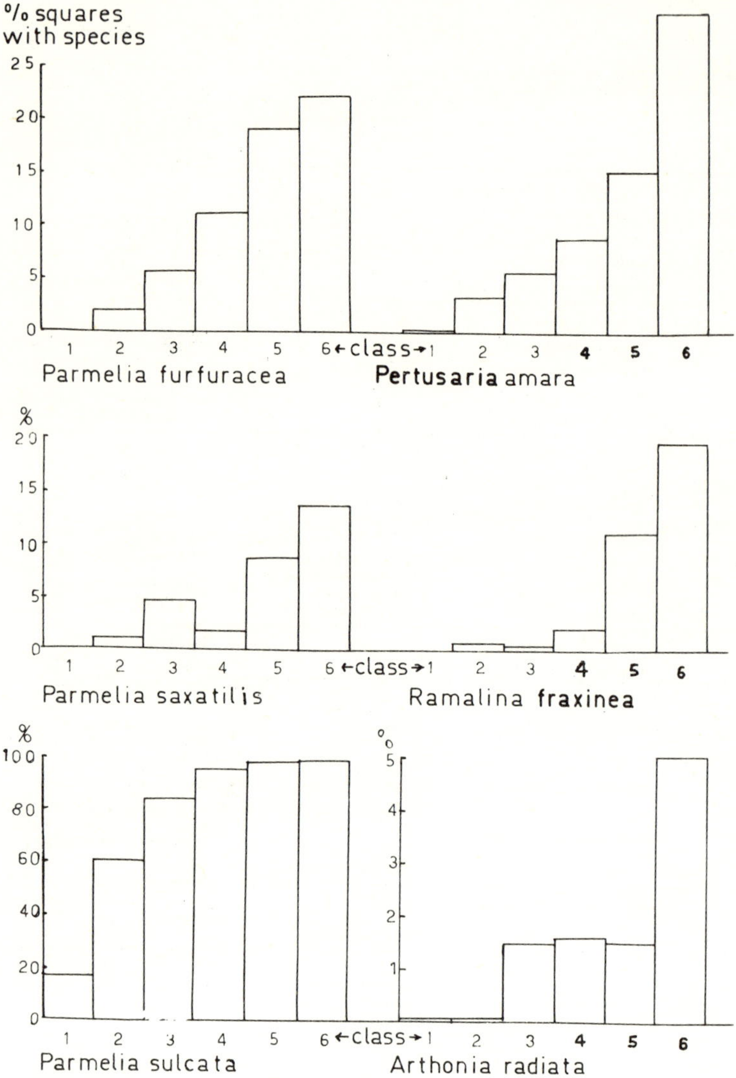

Presence (percentage squares per class with the lichen species) graphs of several lichen species.

appendix 79

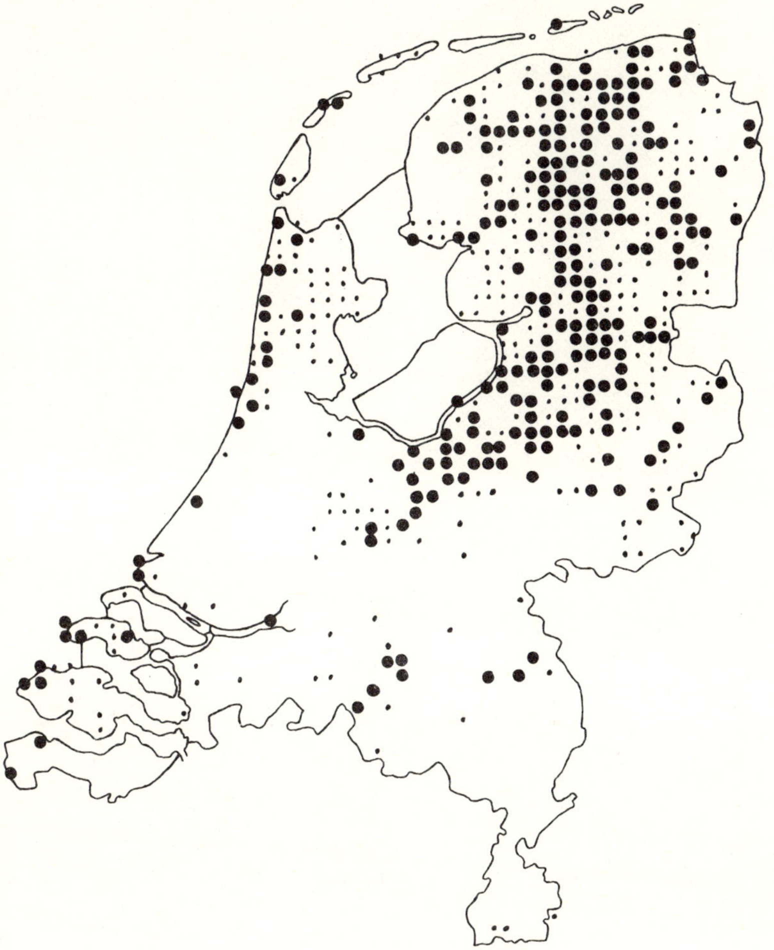

● species belonging to sensitivity class H

. species belonging to sensitivity class G

ACER

FRAXINUS

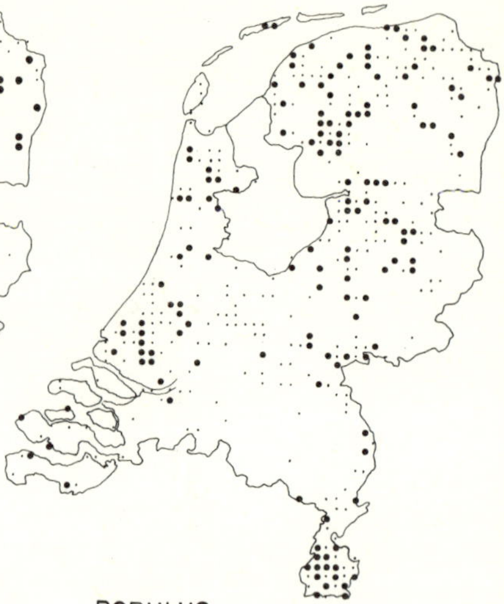

JUGLANS

POPULUS

· treespecies recorded
• treespecies having the highest epiphyte-richness of the square (as compared with other phorophyte species).

QUERCUS

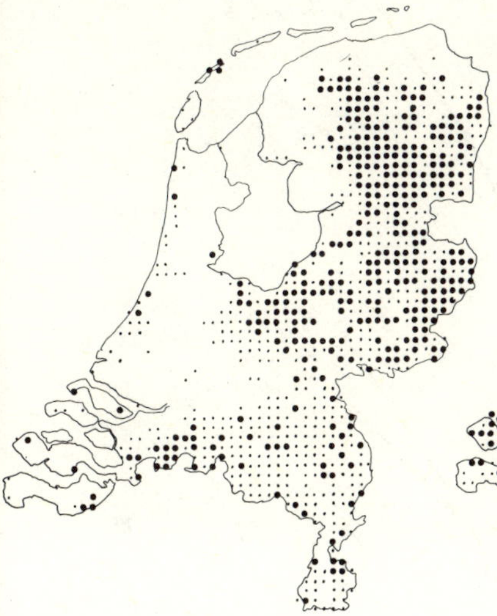

SALIX

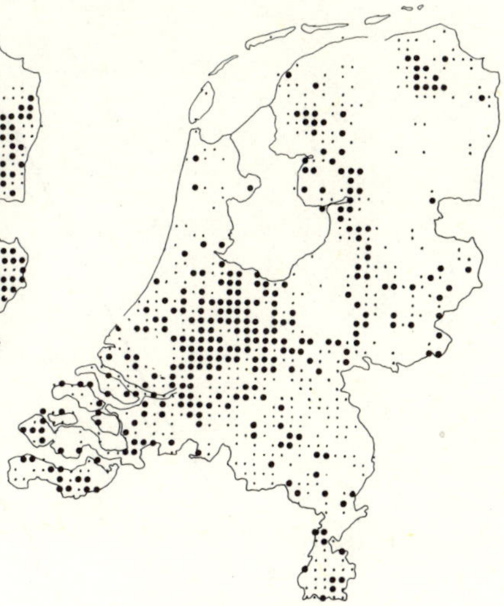

TILIA

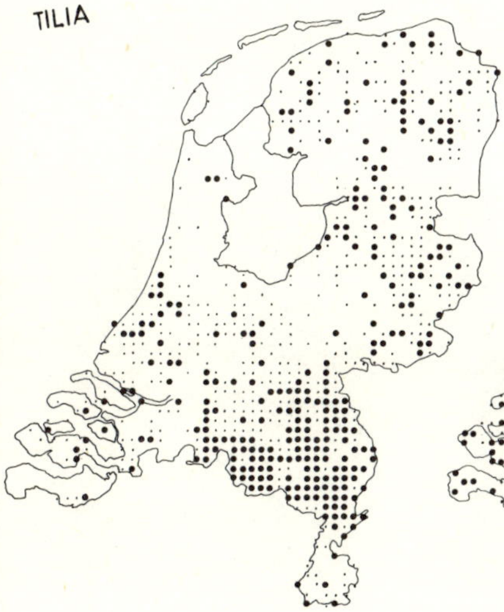

ULMUS

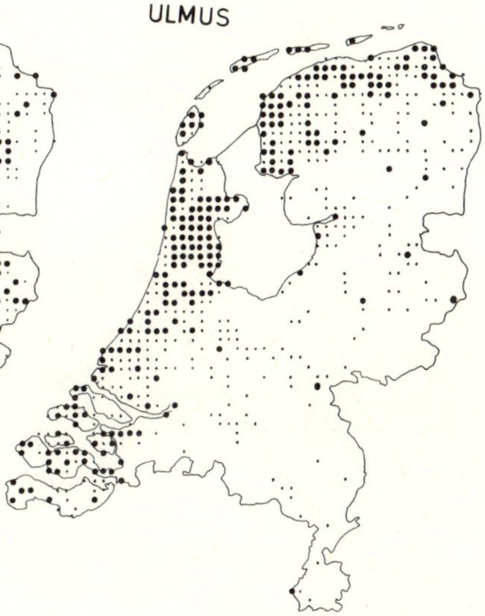

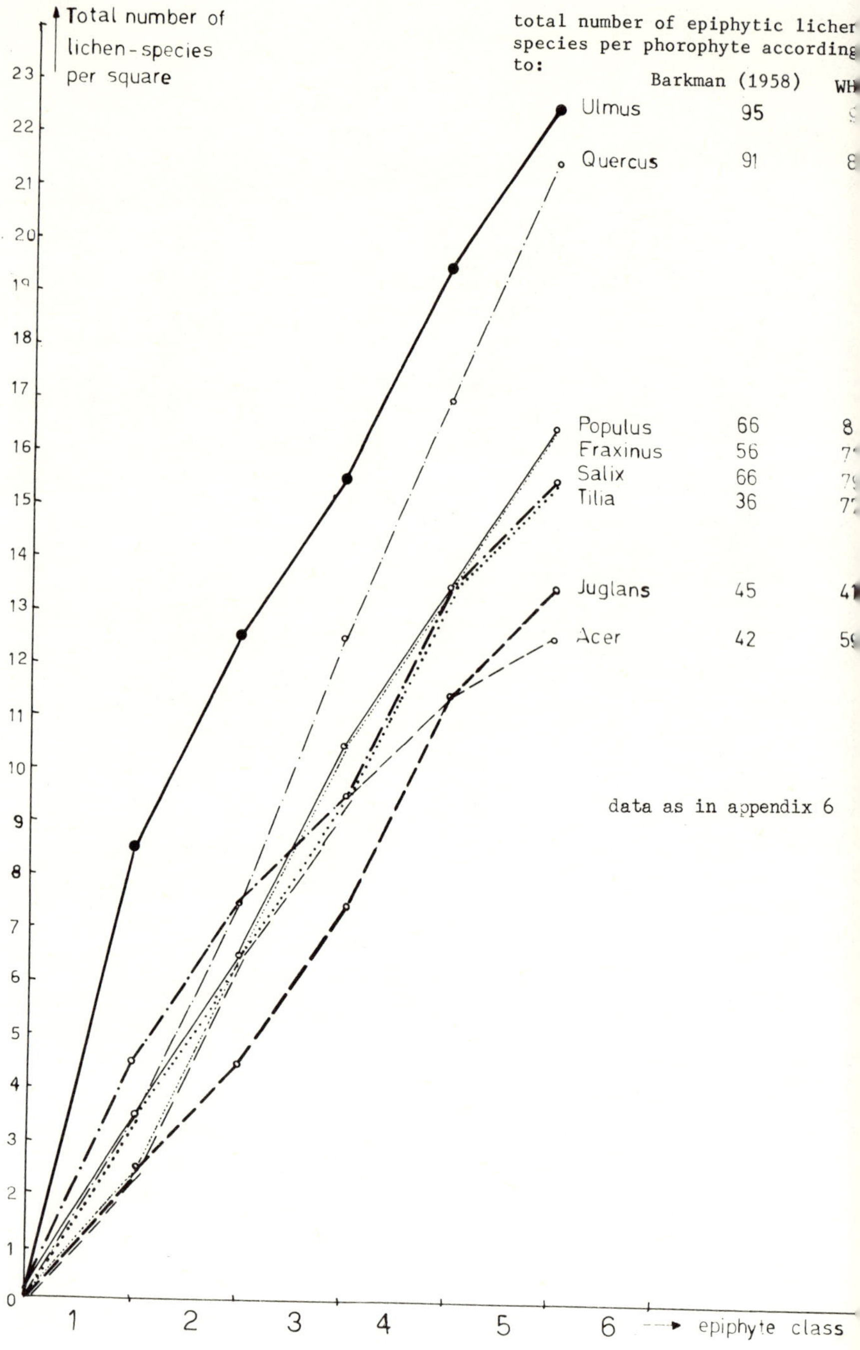
Total number of lichen-species per square
total number of epiphytic licher species per phorophyte according to:
Barkman (1958)
Ulmus 95
Quercus 91
Populus 66
Fraxinus 56
Salix 66
Tilia 36
Juglans 45
Acer 42
data as in appendix 6
epiphyte class
1 2 3 4 5 6
0 1 2 3 4 5 6 7 8 9 10 11 12 13 14 15 16 17 18 20 21 22 23

appendix 83

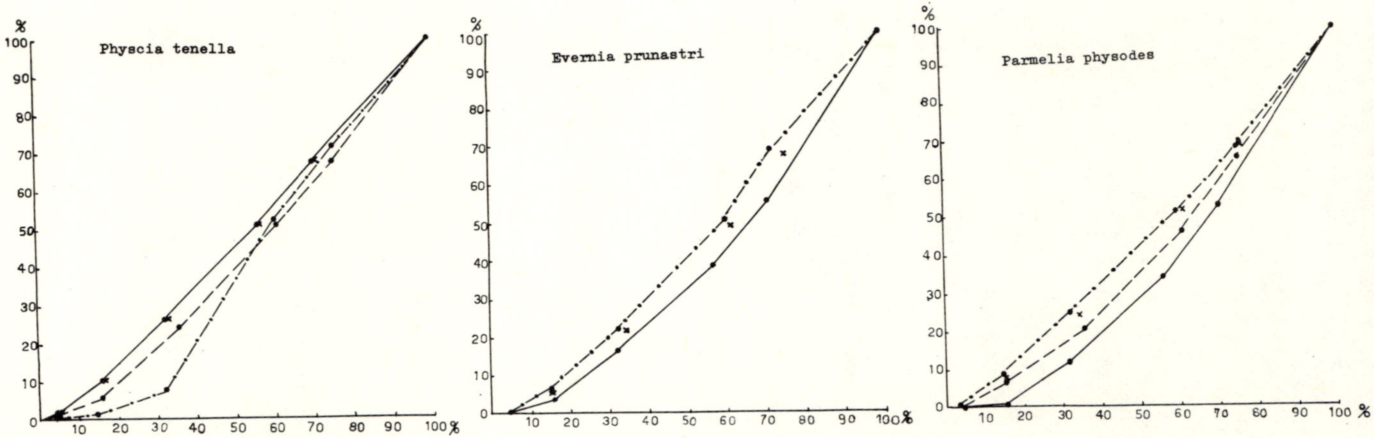

—— on Ulmus
—•— on Quercus
— — on Populus
× total

x-axis : cumulative frequency percentages of squares in which the tree species occurs

y-axis : cumulative frequency percentages of squares in which the lichen species occurs on the indicated tree species

Appendix 84

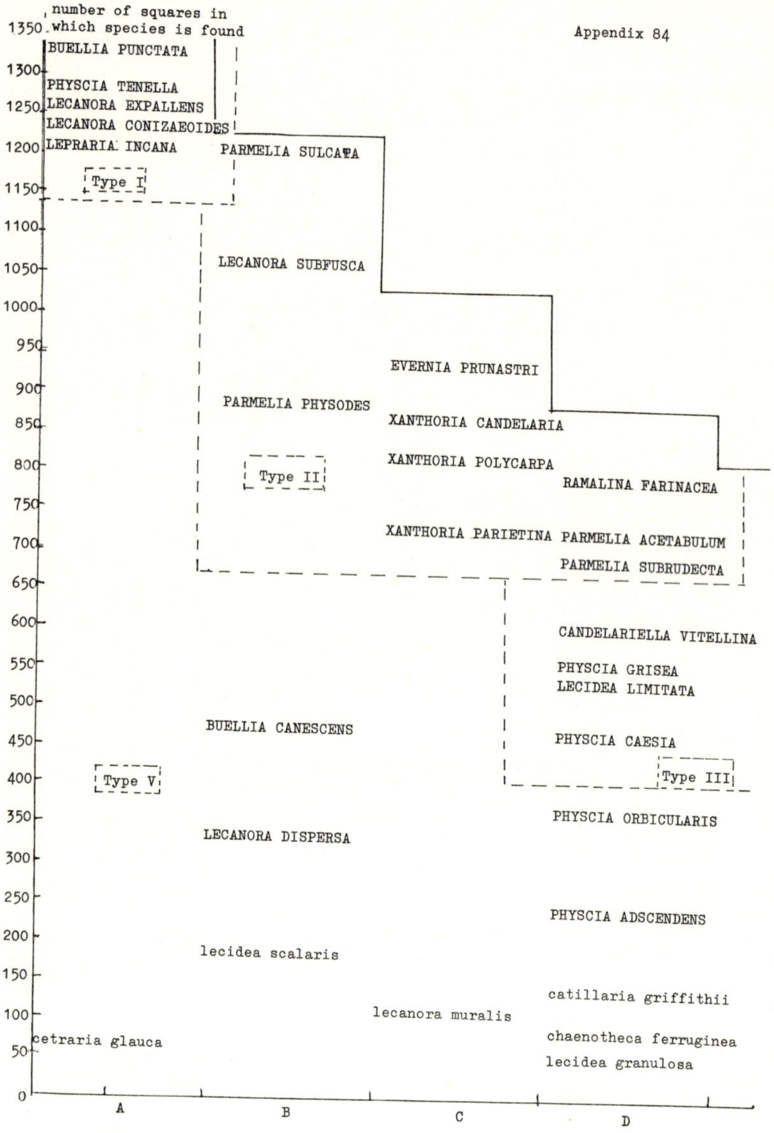
number of squares in which species is found
1350
1300
1250
1200
1150
1100
1050
1000
950
900
850
800
750
700
650
600
550
500
450
400
350
300
250
200
150
100
50
0
BUELLIA PUNCTATA
PHYSCIA TENELLA
LECANORA EXPALLENS
LECANORA CONIZAEOIDES
LEPRARIA INCANA
Type I
PARMELIA SULCATA
LECANORA SUBFUSCA
EVERNIA PRUNASTRI
PARMELIA PHYSODES
XANTHORIA CANDELARIA
XANTHORIA POLYCARPA
Type II
RAMALINA FARINACEA
XANTHORIA PARIETINA
PARMELIA ACETABULUM
PARMELIA SUBRUDECTA
CANDELARIELLA VITELLINA
PHYSCIA GRISEA
LECIDEA LIMITATA
BUELLIA CANESCENS
PHYSCIA CAESIA
Type V
Type III
PHYSCIA ORBICULARIS
LECANORA DISPERSA
PHYSCIA ADSCENDENS
lecidea scalaris
catillaria griffithii
lecanora muralis
cetraria glauca
chaenotheca ferruginea
lecidea granulosa
A
B
C
D

Appendix 84

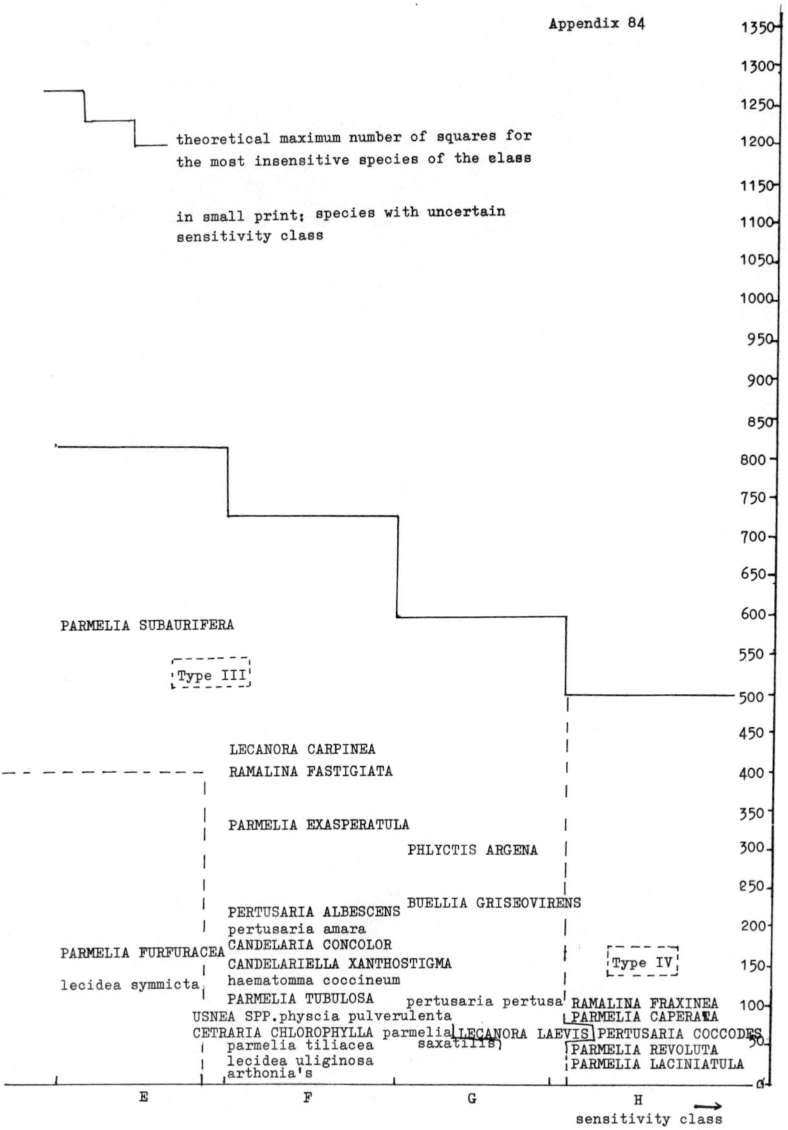

theoretical maximum number of squares for the most insensitive species of the class
in small print: species with uncertain sensitivity class
1350
1300
1250
1200
1150
1100
1050
1000
950
900
850
800
750
700
650
600
550
500
450
400
350
300
250
200
150
100
50
0
PARMELIA SUBAURIFERA
Type III
LECANORA CARPINEA
RAMALINA FASTIGIATA
PARMELIA EXASPERATULA
PHLYCTIS ARGENA
BUELLIA GRISEOVIRENS
PERTUSARIA ALBESCENS
pertusaria amara
PARMELIA FURFURACEA
CANDELARIA CONCOLOR
CANDELARIELLA XANTHOSTIGMA
lecidea symmicta
haematomma coccineum
PARMELIA TUBULOSA
pertusaria pertusa
USNEA SPP.
physcia pulverulenta
CETRARIA CHLOROPHYLLA
parmelia saxatilis
LECANORA LAEVIS
parmelia tiliacea
lecidea uliginosa
arthonia's
Type IV
RAMALINA FRAXINEA
PARMELIA CAPERATA
PERTUSARIA COCCODES
PARMELIA REVOLUTA
PARMELIA LACINIATULA
E
F
G
H
sensitivity class

appendix 85

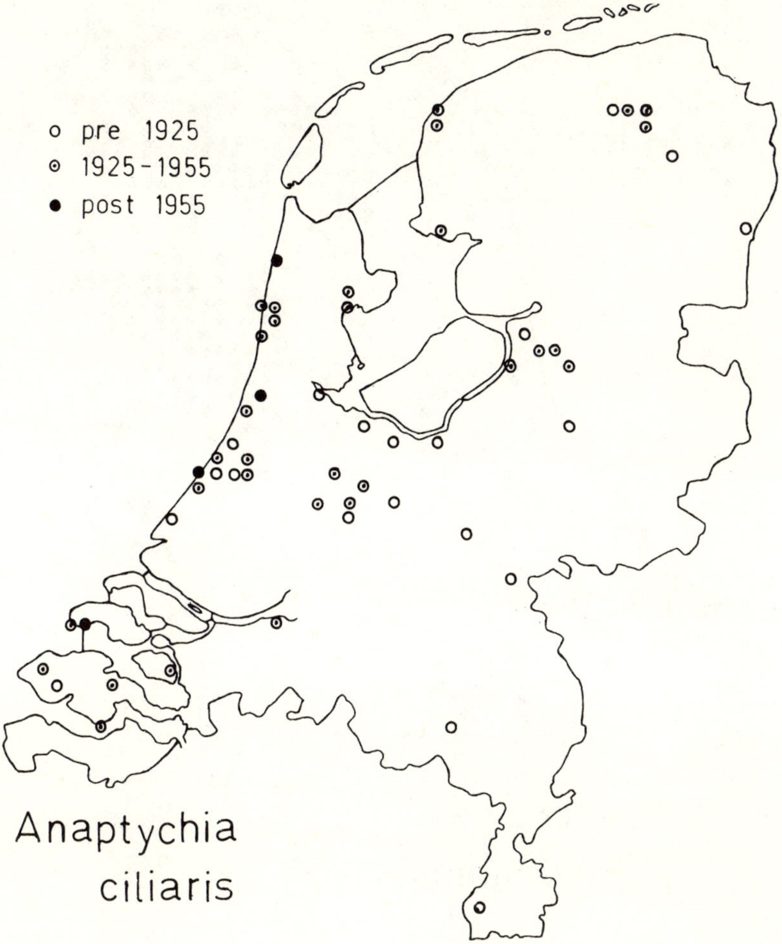

Anaptychia ciliaris

appendix 86

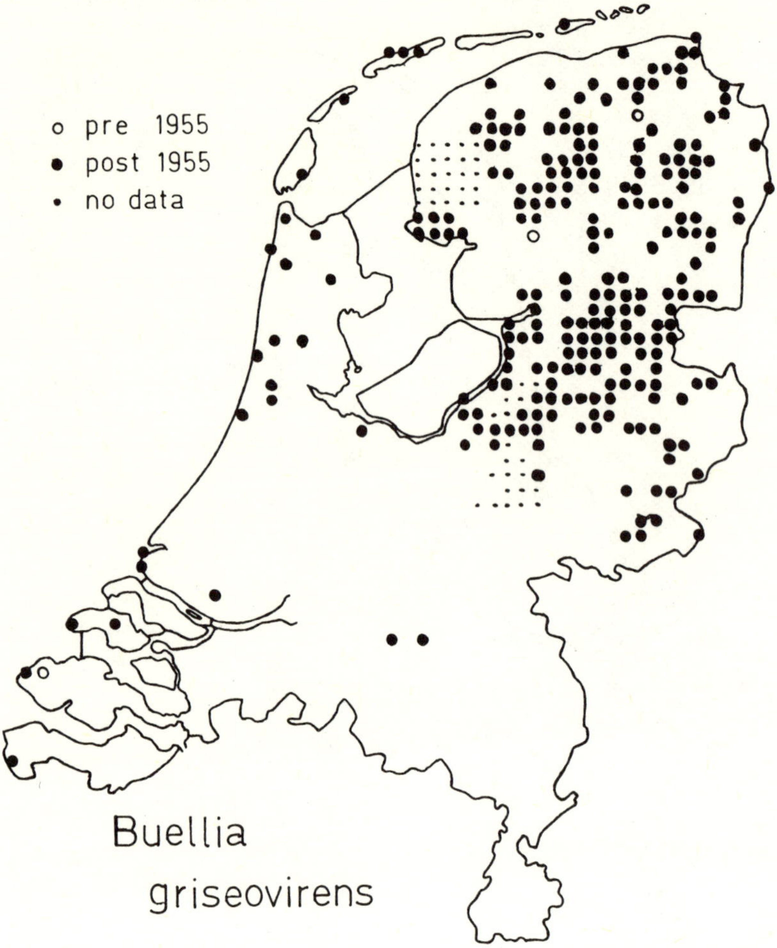

appendix 87

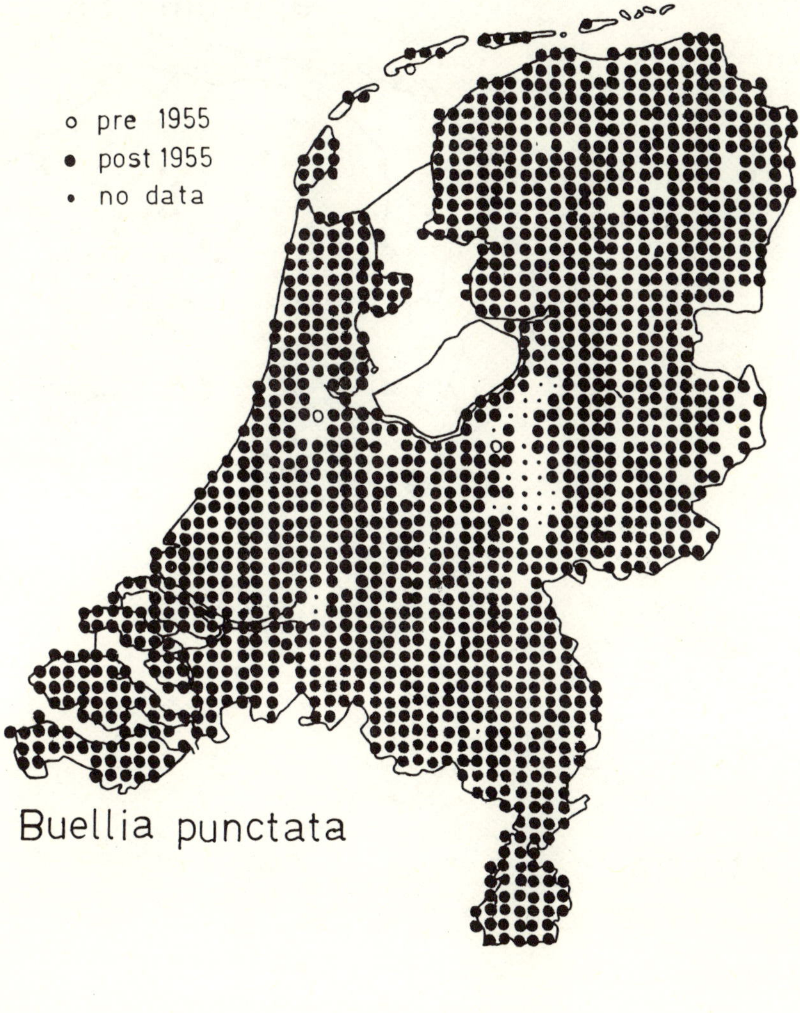

Buellia punctata

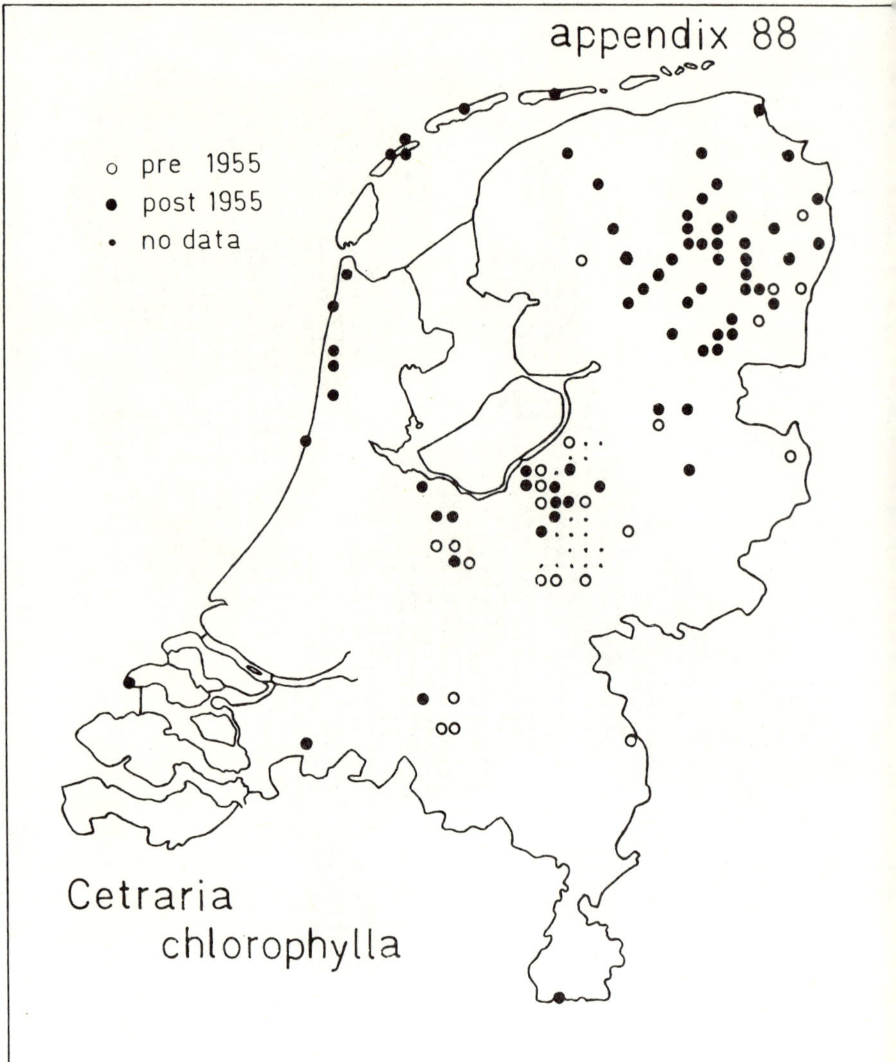

appendix 88
o pre 1955
● post 1955
• no data
Cetraria
chlorophylla

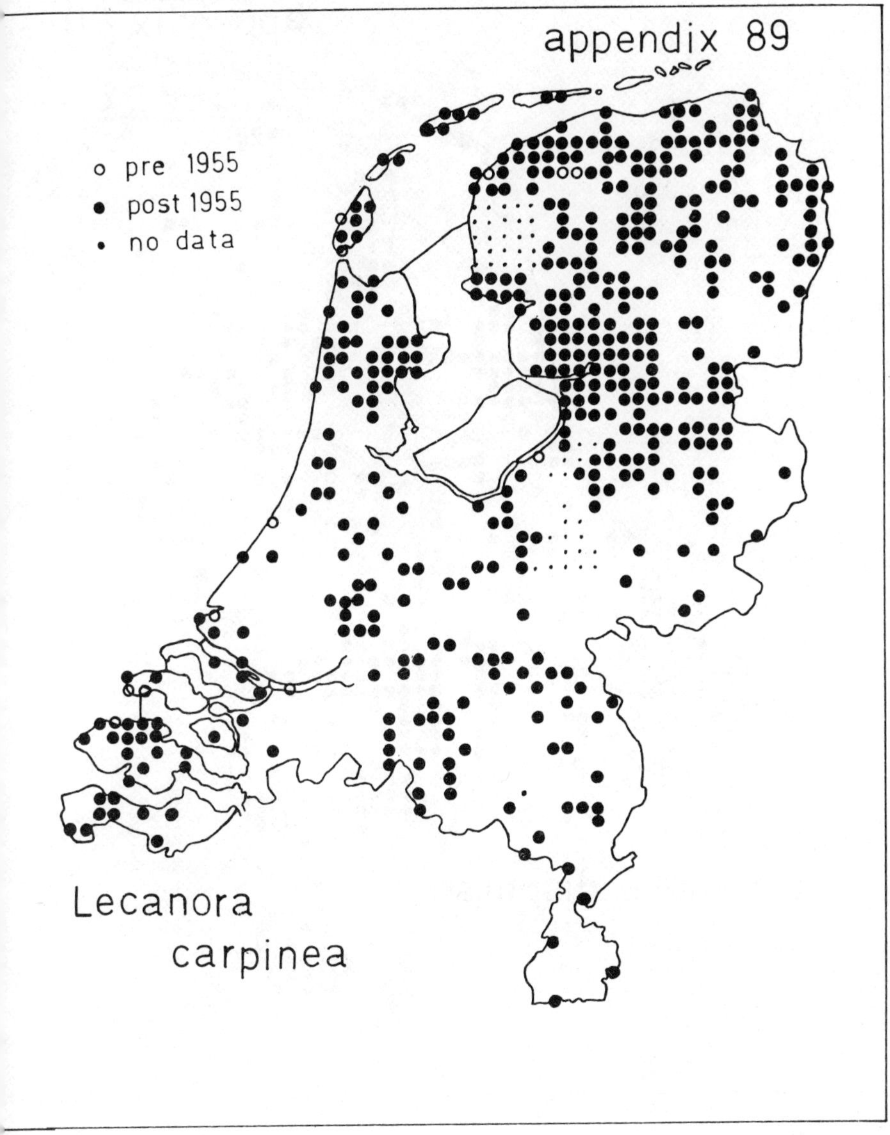
appendix 89
o pre 1955
● post 1955
• no data
Lecanora
carpinea

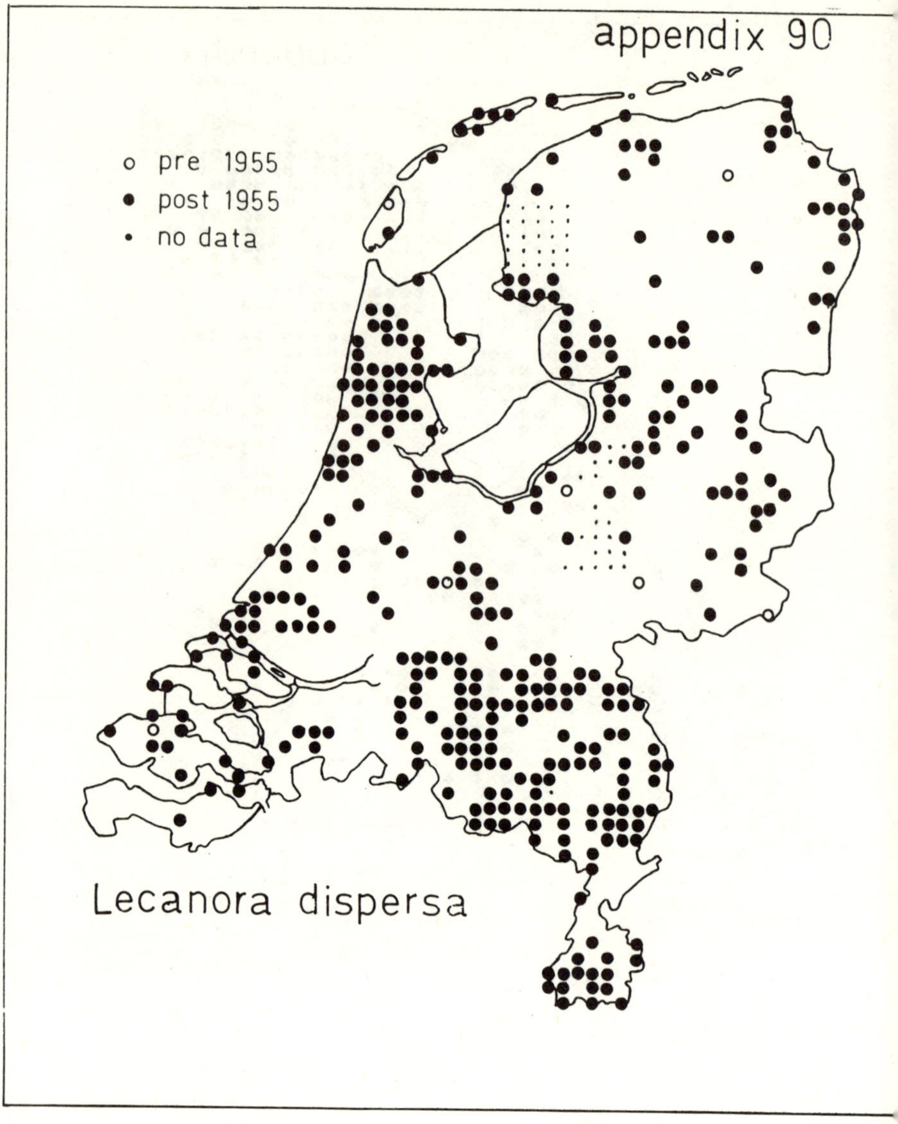
appendix 90
o pre 1955
• post 1955
· no data
Lecanora dispersa

appendix 91

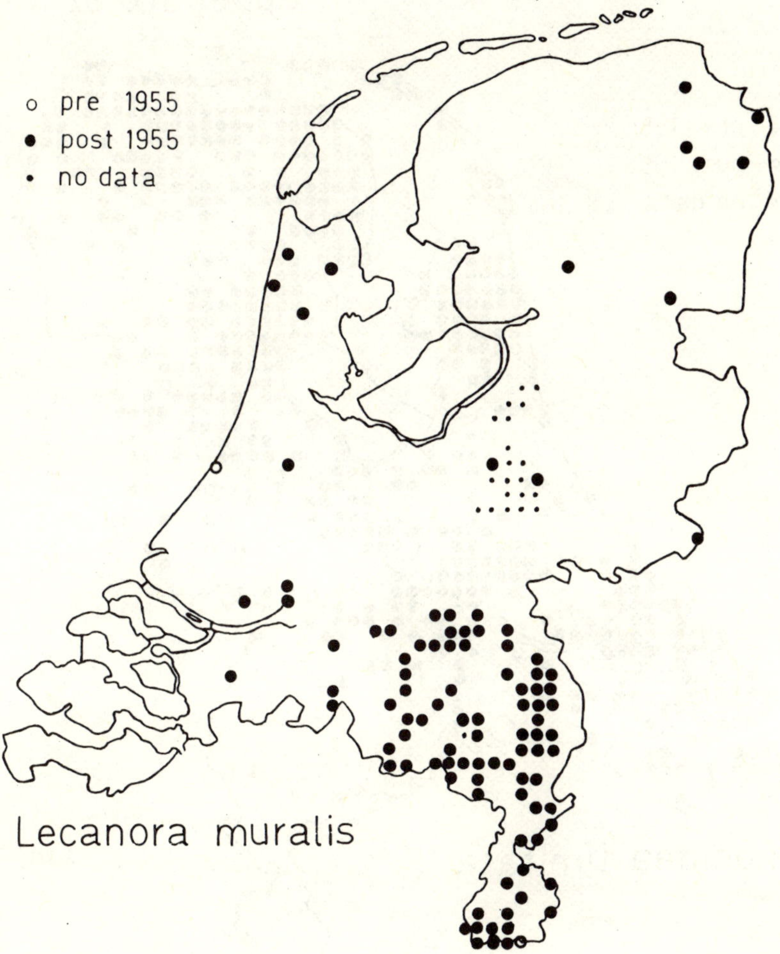

Lecanora muralis

appendix 92

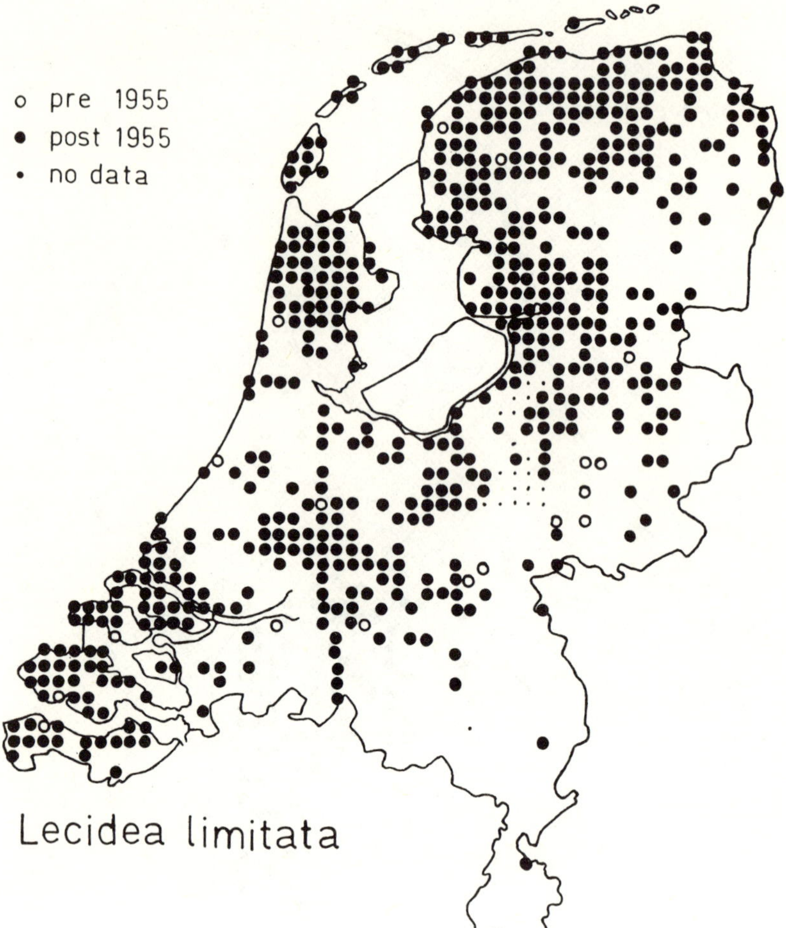

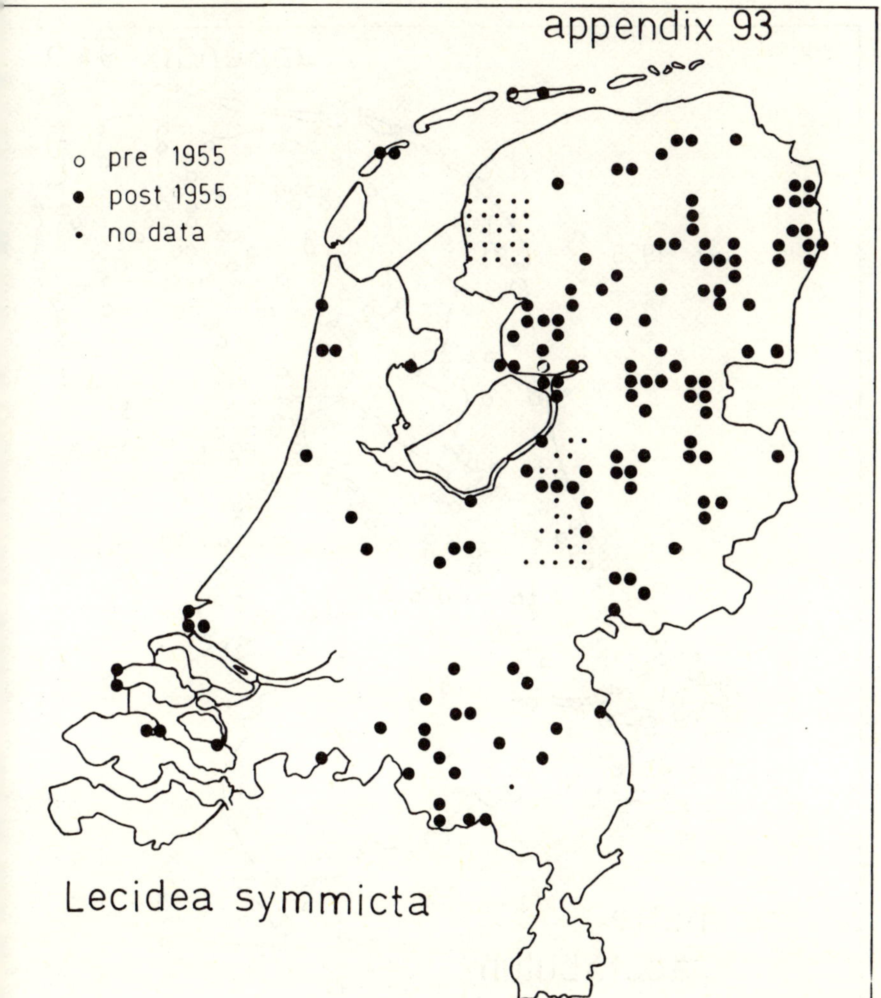
appendix 93
o pre 1955
● post 1955
• no data
Lecidea symmicta

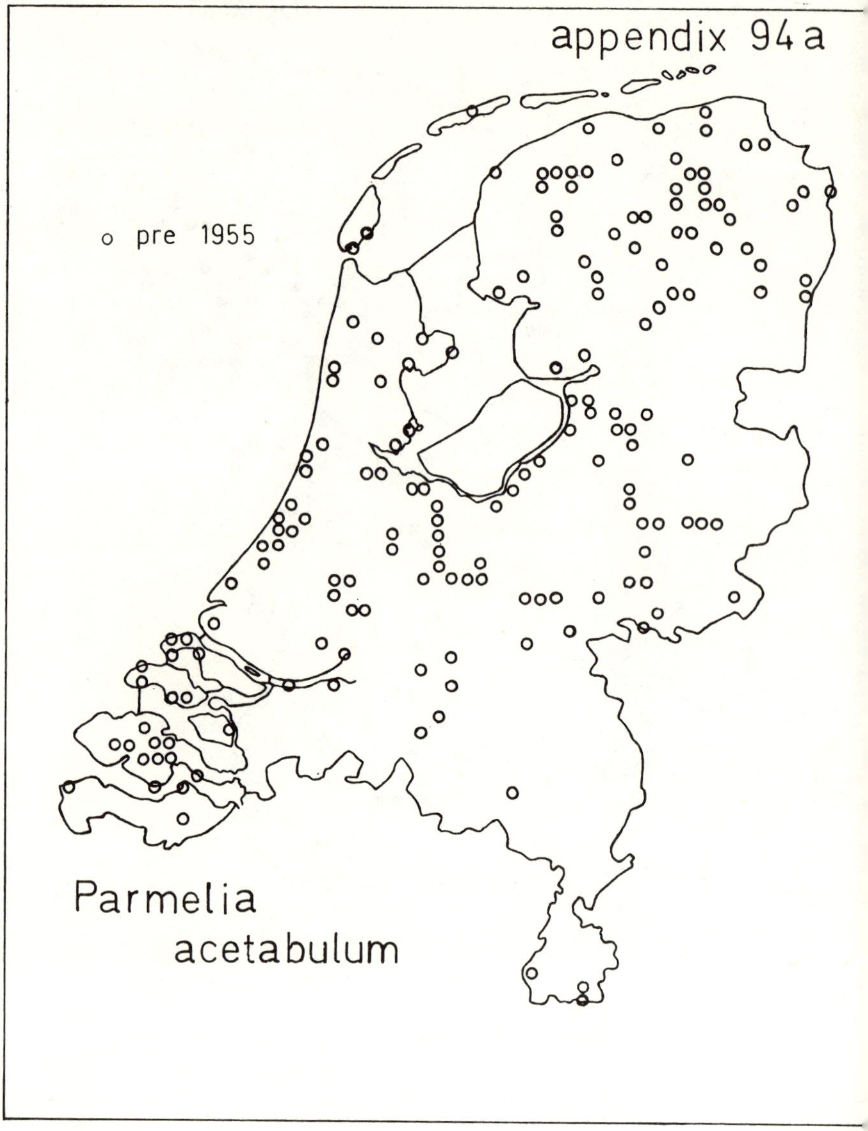
appendix 94a
o pre 1955
Parmelia
acetabulum

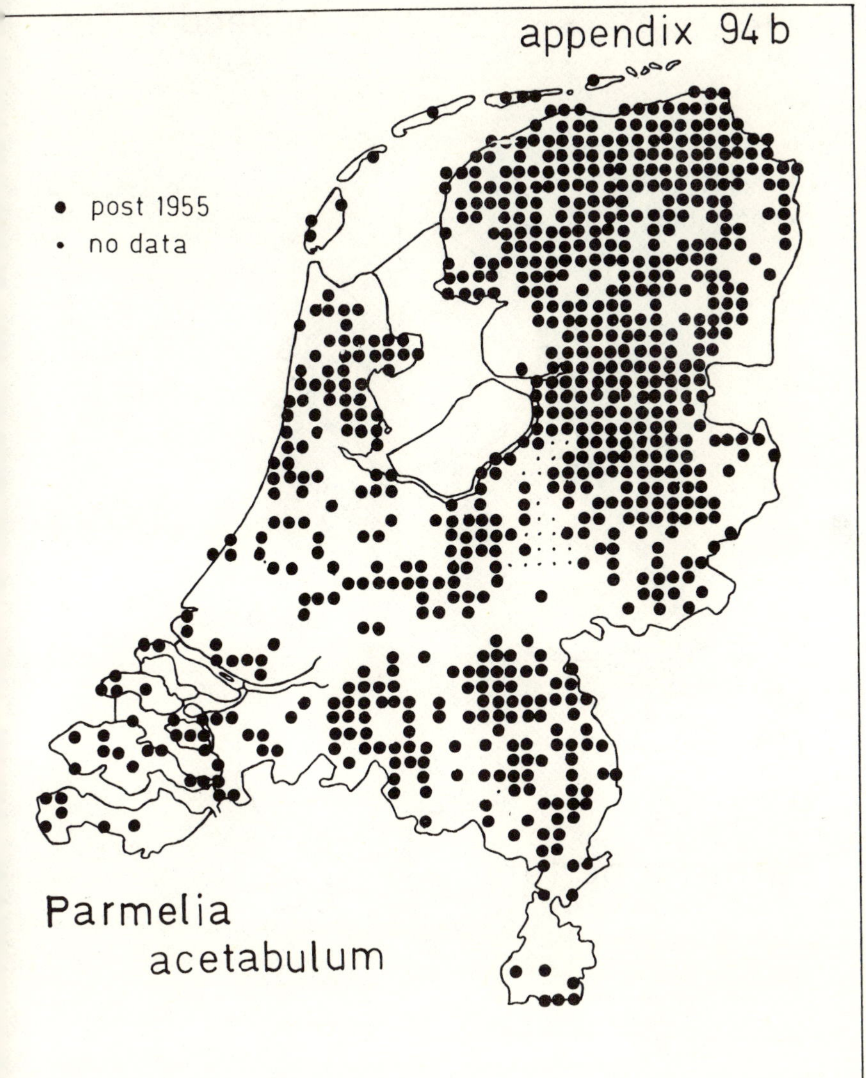
appendix 94 b
post 1955
no data
Parmelia
acetabulum

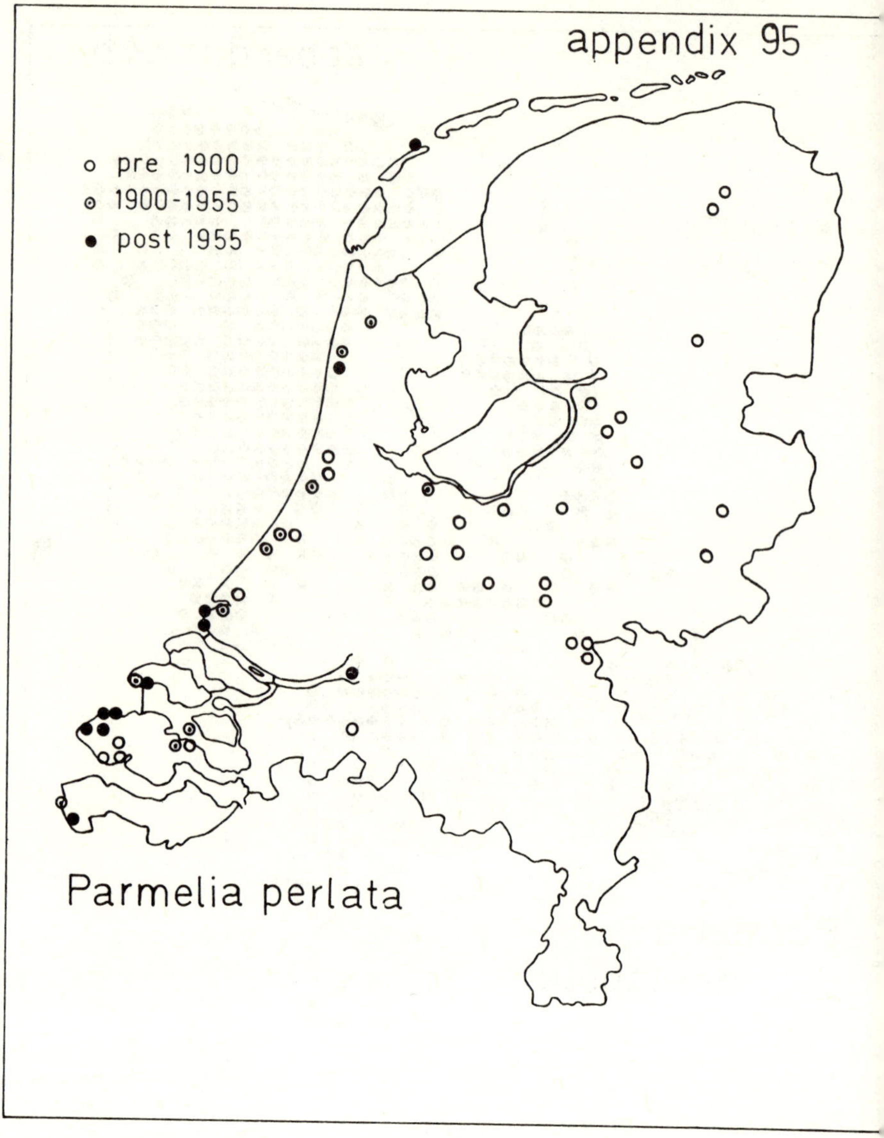
appendix 95
o pre 1900
◎ 1900-1955
● post 1955
Parmelia perlata

appendix 96

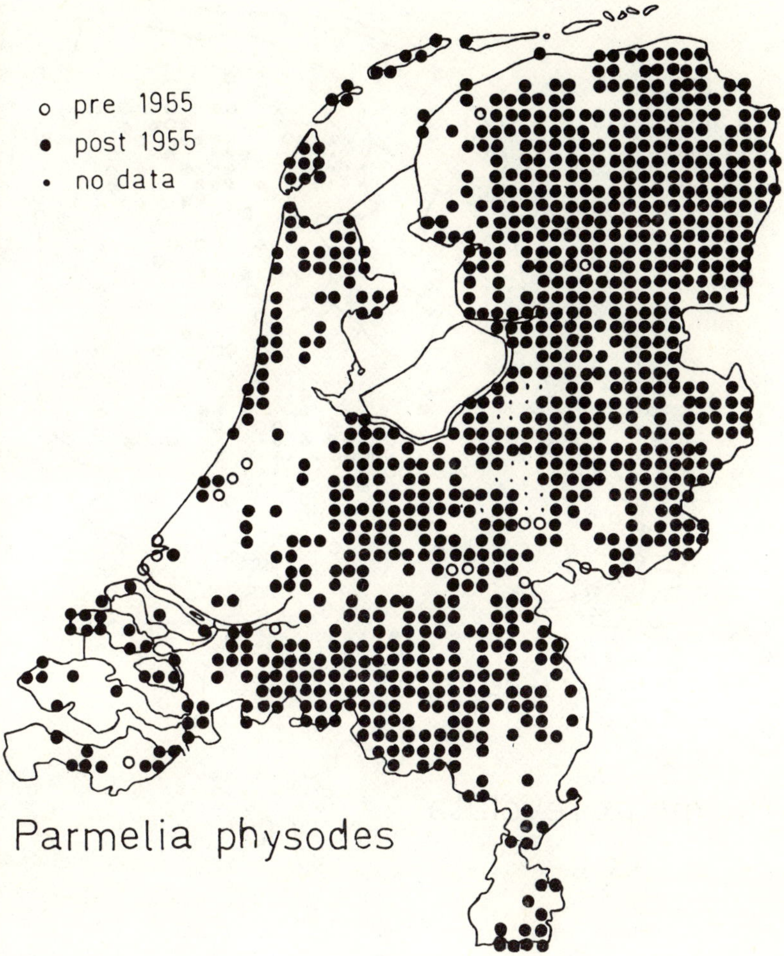

Parmelia physodes

appendix 97

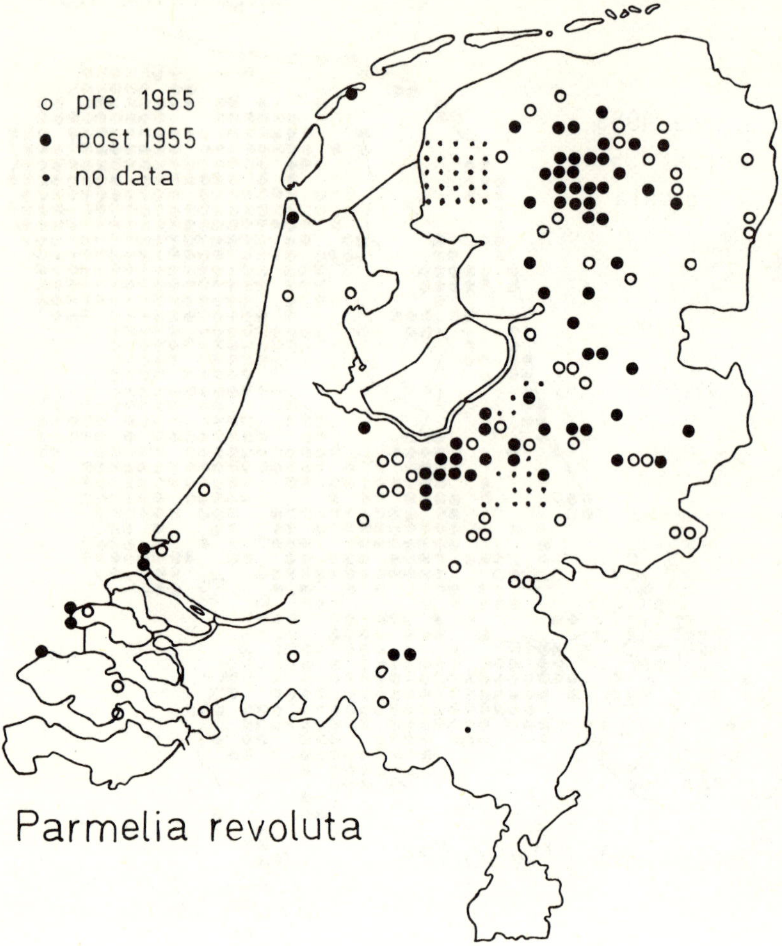

Parmelia revoluta

appendix 98

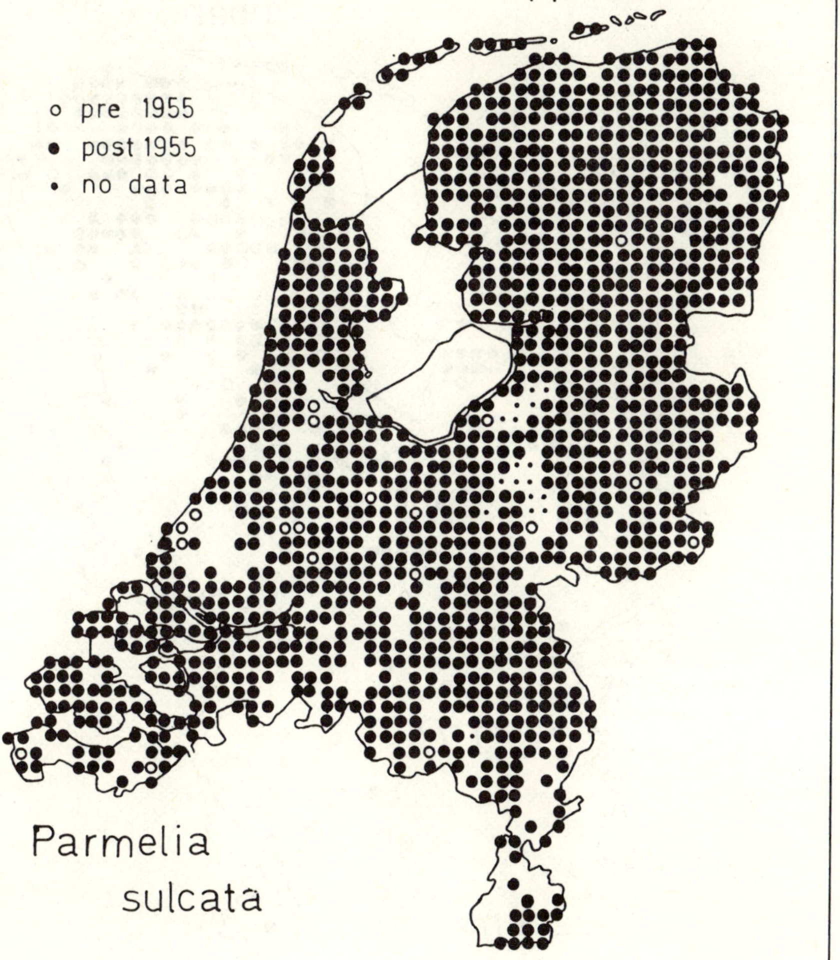

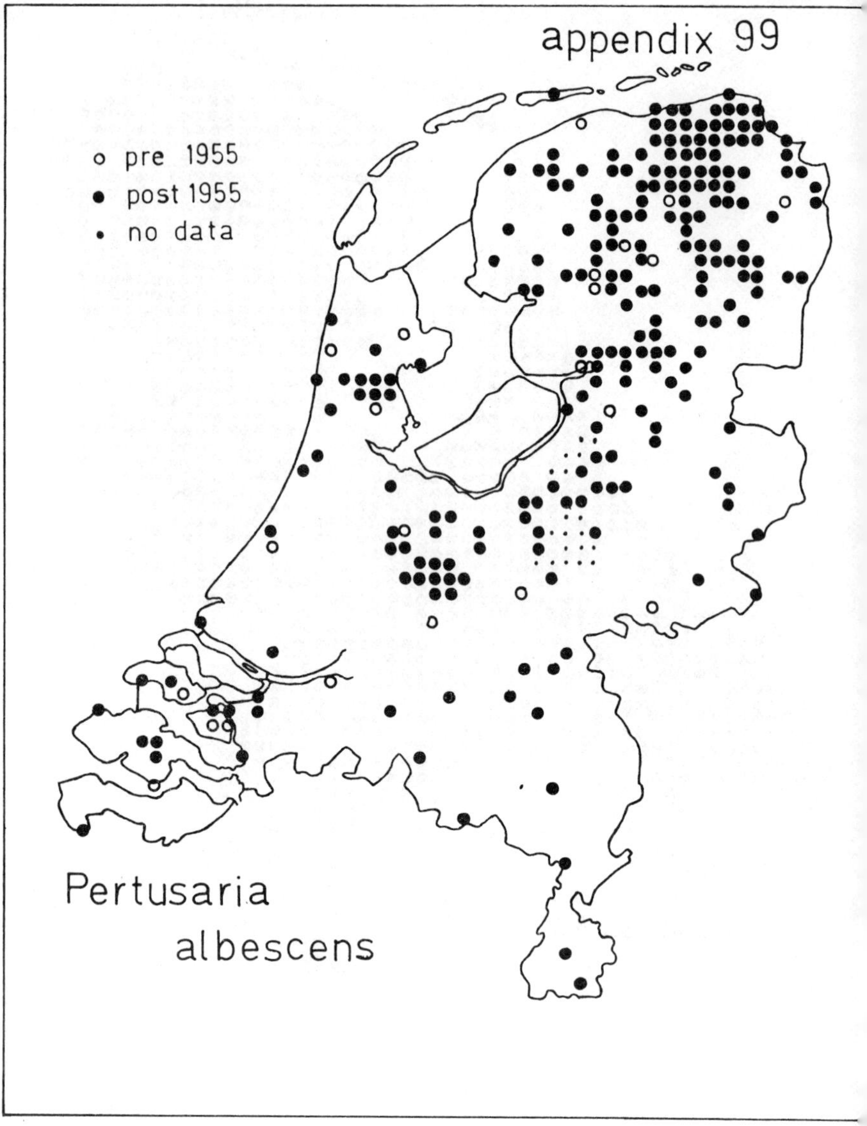
appendix 99
o pre 1955
● post 1955
• no data
Pertusaria
albescens

appendix 100

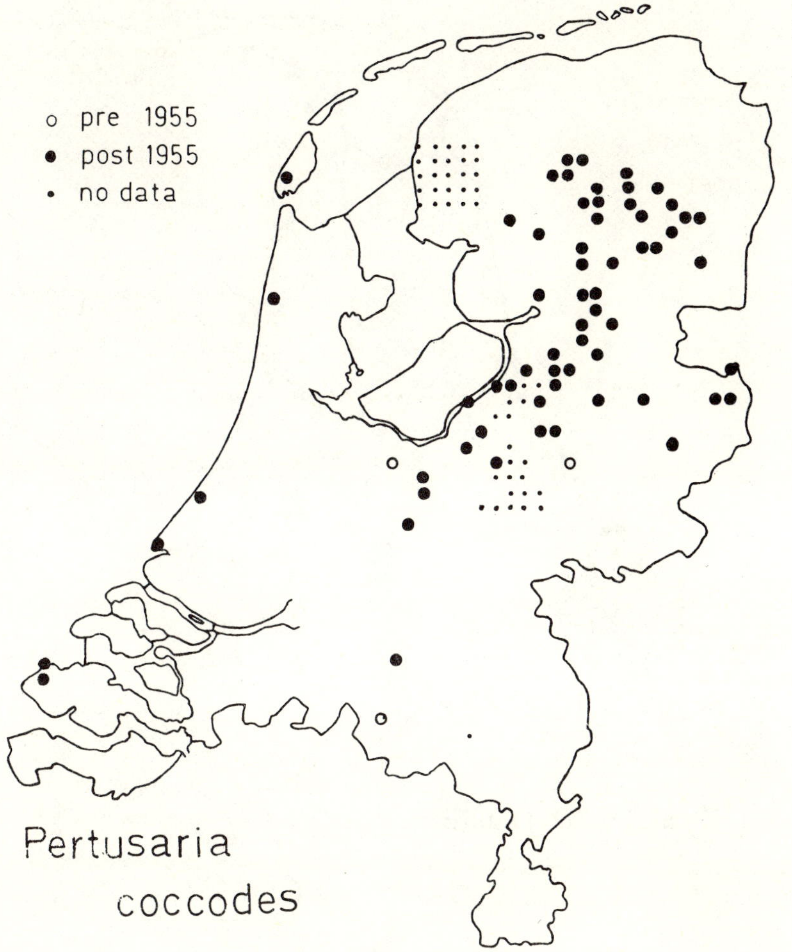

appendix 101

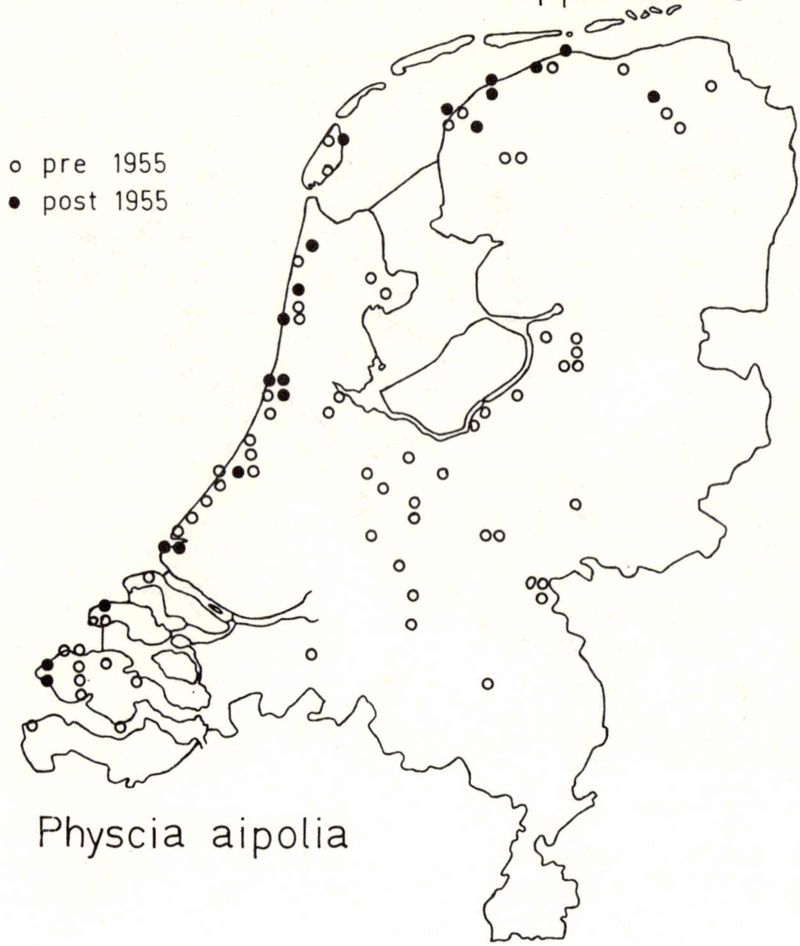

Physcia aipolia

appendix 102

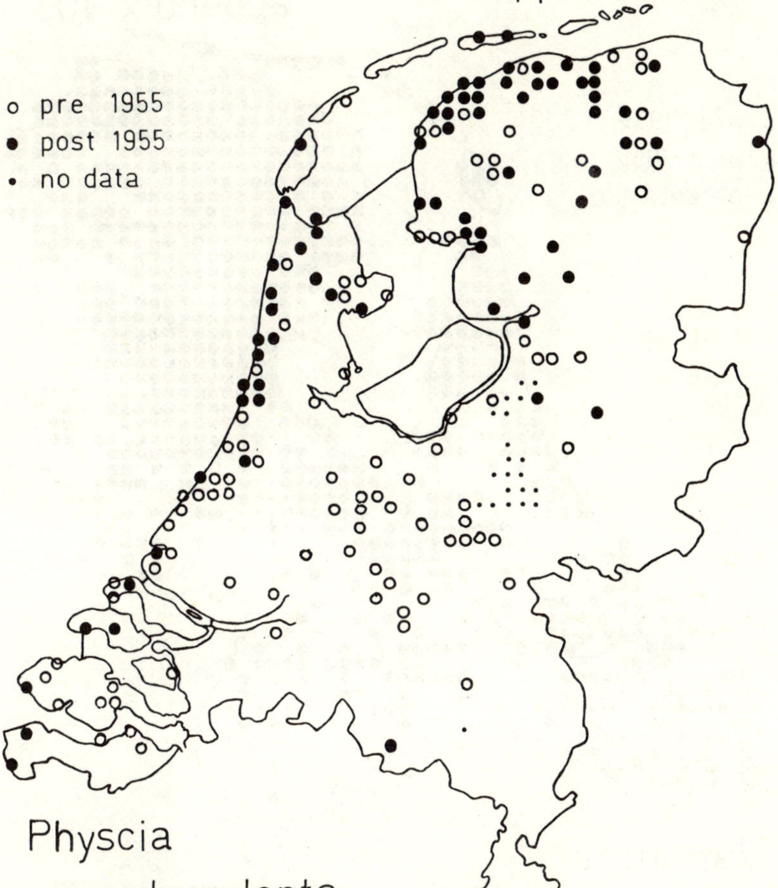

Physcia pulverulenta

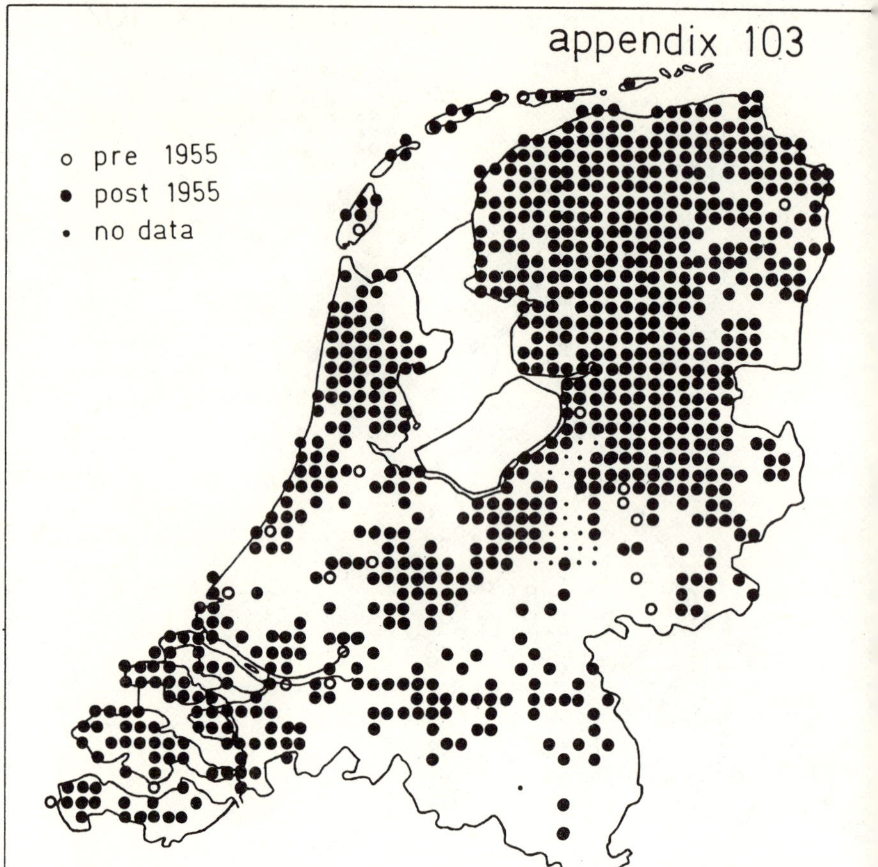
appendix 103
o pre 1955
● post 1955
• no data
Ramalina
farinacea

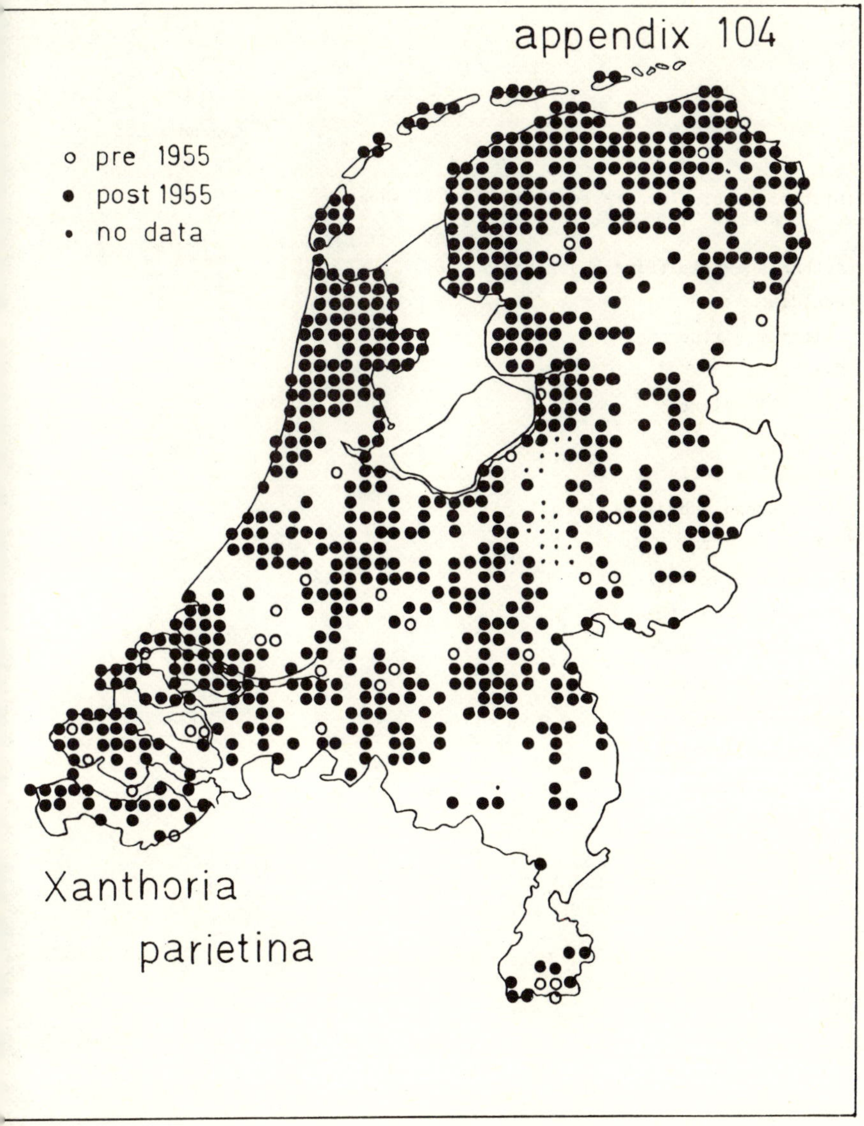
appendix 104
o pre 1955
• post 1955
• no data
Xanthoria
parietina

List of species/genera found in 10 - 25 squares

Arthonia exilis (Floerke) Sandst. c.f.

Bacidia

Calicium viride

Caloplaca

Dimerella diluta

Graphis

Lecania cyrtella

Ochrolechia androgyna

Opegrapha atra

Parmelia elegantula

Parmelia glabratula

Parmeliopsis ambigua

Pertusaria leioplaca

Physcia aipolia

Porina chlorotica

Rinodina exigua

appendix 106

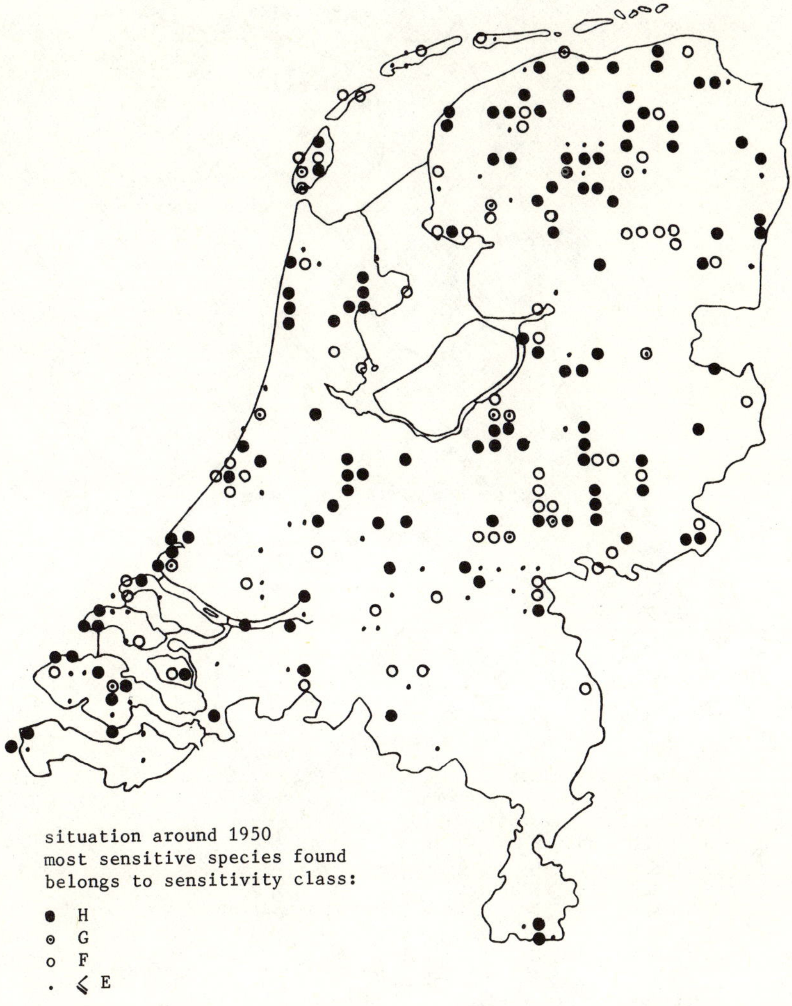

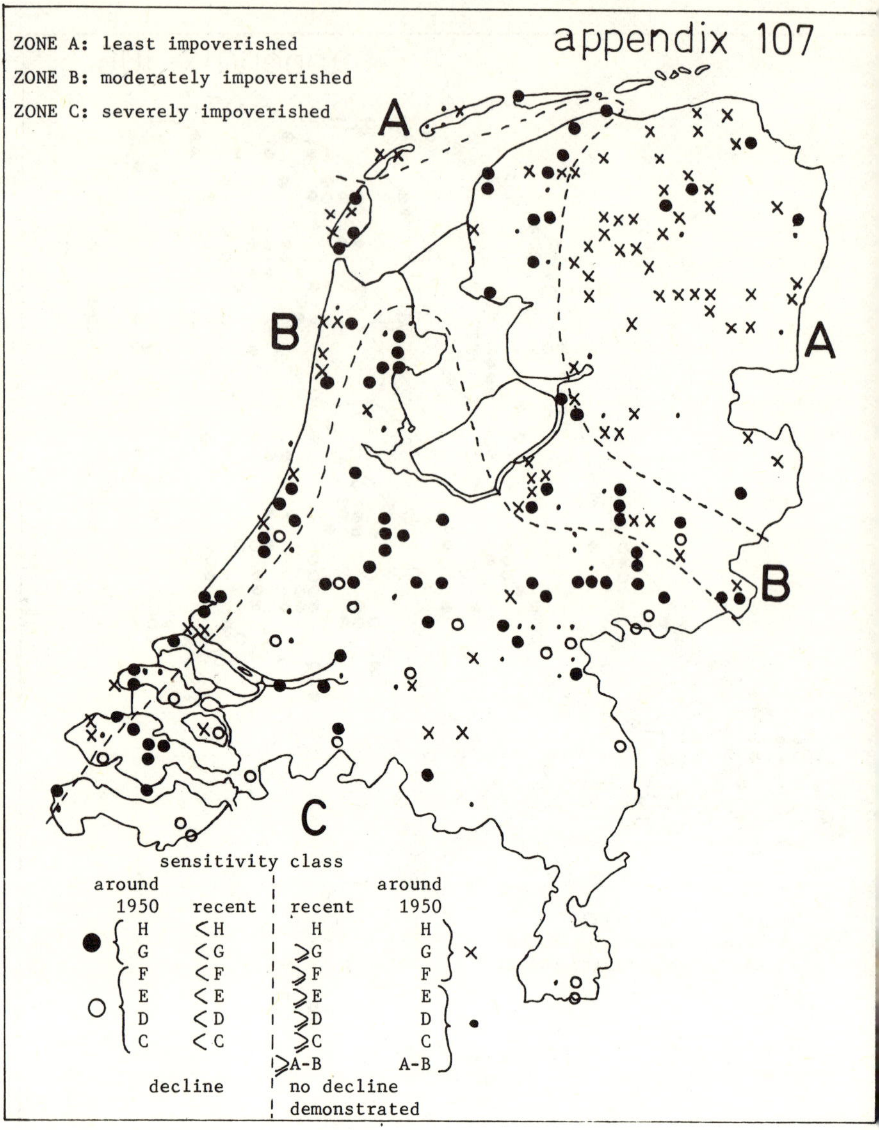
appendix 107
ZONE A: least impoverished
ZONE B: moderately impoverished
ZONE C: severely impoverished
A
B
A
B
C
sensitivity class
around 1950 | recent | recent | around 1950
● H | <H | H | H ×
● G | <G | ≥G | G ×
○ F | <F | ≥F | F ×
○ E | <E | ≥E | E •
○ D | <D | ≥D | D •
○ C | <C | ≥C | C •
≥A-B | A-B •
decline
no decline demonstrated

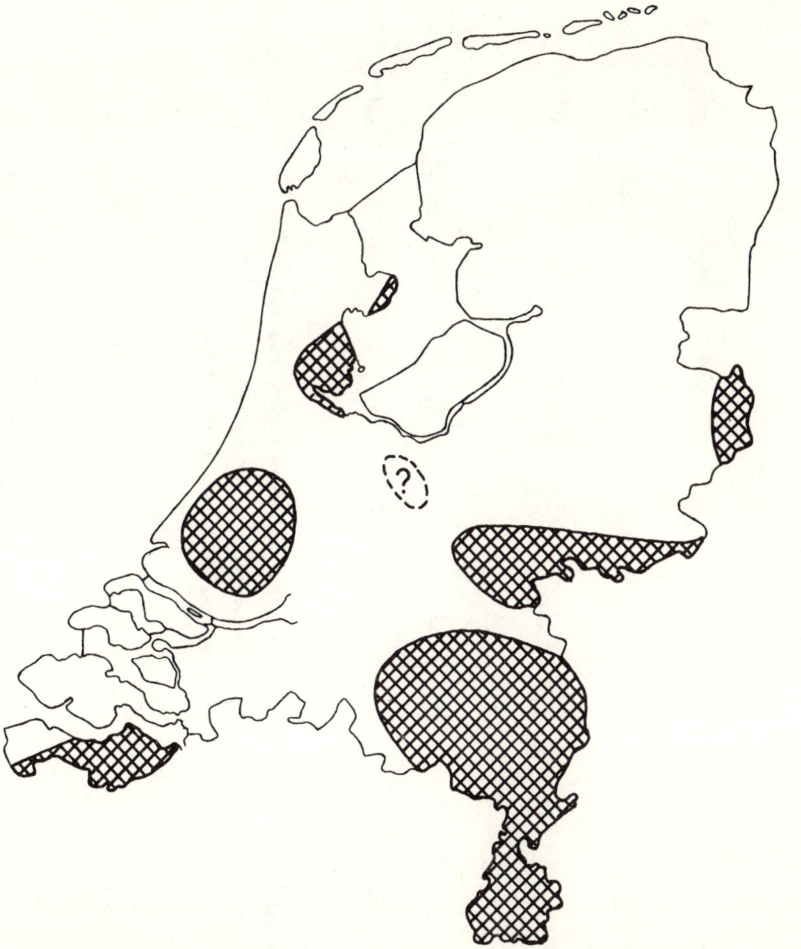

Probable size of „epiphyte-deserts", ca. 1950; an epiphyte-desert is defined as an area belonging approximately to classes 1 - 3 in wich no H or G species occur.
The map is based on Appendix 1 and Appendix **106**

HF Fumigations

	experiment 1		experiment 2	
	[HF] low	[HF] high	[HF] low	[HF] high
control glasshouse	0.0 - 0.20 - 0.5	0.10 - 0.2 - 0.4	0.10 - 0.18 - 0.24	
HF glasshouse	0.3 - 0.46 - 1.0	1.4 - 2.1 - 3.7	0.18 - 0.26 - 0.44	1.0 - 2.2- 3.4
temperature (^{o}C)	5 - 10 - 20 - 30	5 - 16 - 28	5 - 10 - 15 - 18	
relative humidity (%)	40- 60 - 80 - 95	40 - 65 - 95	45 - 60 - 80 - 90	
period	May 10 - Sept. 29 1972 137 days	May 10 - July 1972 75 days	January 12 - April 6 1973 84 days	

Concentrations are given in ppb. Underlined values are total averages (where two values are underlined, these represent the lowest and highest day averages). The values to the left and right of the underlined values are the lowest and highest values measured, respectively.

Appendix 110

gas	fumigation experiment	Control Exp. 3	Control Exp. 4	C_2H_4 Exp. 3	SO_2 Exp. 3	SO_2 Exp. 4	O_3 Exp. 3	O_3 Exp. 4	SO_2 + O_3 Exp. 4
C_2H_4	day (12hr)			112					
	night			-					
SO_2	average (24hr)	ca. 5	ca. 5		49.9	57.3	ca. 5	ca. 5	55.3
	day (12hr)				93				
	day (8hr)					142			138
	night (12hr)				24				
	night (16hr)					15			14
O_3	average (24hr)	ca. 16	20.6		ca. 16	19.7	25.5	29	32
	day (12hr)	ca. 25			ca. 25		44		
	day (8hr)		30			31		59	64
	night (12hr)	7			7		7		
	night (16hr)		16			14		14	16

	experiment 3	experiment 4
temperature (^{o}C)	3 - <u>6</u> - <u>22</u> - 30	<u>10</u> - <u>20</u>
relative humidity (%)	40 - <u>50</u> - <u>85</u> - 95	<u>70</u> - <u>80</u>
period	May 14 - October 31 1973 170 days	May 9 - November 4 1974 180 days

Concentrations are given in ppb. Estimated values are indicated by <u>ca</u>. For the temperature and relative humidity underlined values are lowest and highest day averages; for experiment 3 the absolute minima and maxima are additionally given.

Number of days	forest	number of days exposure to gas	forest	control	[HF] low	control	[HF] high	number of days exposure to gas	forest
Cetraria glauca									
75	0- 0 - 0	75	-	0- 2.6- 5	0- 5.4- 15	0- 2.6- 5	2 - [7.5]- 15	75	-
137	5- 28.8- 60	137	-	0-77.0-100	5-55.0- 90	5-22.5- 60	20 -31.4- 50	75	62
241	5- 35.0- 90	137	104	70-92.5-100	10-57.5- 95	5-31.5-100	20 -52.6-100	75	166
Evernia prunastri									
75	10- 24.4- 50	75	-	0- 8.8- 20	0- 7.6- 20	0- 8.8- 20	0 - 2.8- 10	75	-
137	85- 96.3-100	137	-	30-37.5- 60	10-[52.5]- 90	5-10.0- 20	0 -[73.6]-100	75	62
241	100-100 -100	137	104	30-45.0- 60	5-[77.5]-100	10-77.5-100	50 -91.4-100	75	166
Parmelia physodes									
75	0- 5.6- 15	75	-	0-53 - 30	0- 9.0- 60	0- 5.3- 30	7.5-[32.0]- 70	75	-
137	0- 5.6- 15	137	-	15-65.0- 95	10-65.0-100	0- 7.3- 30	5.0-[47.6]-100	75	62
241	0- 5.8- 15	137	104	50-81.0- 95	10-66.3-100	0- 9.7- 40	5.0-[59.5]-100	75	166
Parmelia saxatilis									
75	0- 0 - 0	75	-	5-12.8- 50	0-24.4- 80	5-12.8- 50	5 -[32.2]- 90	75	-
137	0- 19.4- 60	137	-	5-57.0- 98	5-53.1-100	10-55.0-100	10 -48.6-100	75	62
241	0- 29.4-100	137	104	50-74.0- 98	5-55.0-100	50-65.0-100	10 -75.0-100	75	166
Number of samples per species		4		4 (8)	7-8	4 (8)	7-8		

Experiment 1. In each column lowest, average and highest percentages discoloured surface area are given. Boxed average values indicate significant divergence from the control values. The 75-day control values are those of all 8 control samples. After the 75th day, half of the controls were transfered to the forest with the high-HF samples.

	"forest"	"field"	control	[HF] very low	[HF] high	number of samples
Centraria glauca						
56 days dry	0-4-20	10-50	0-14.6- 60	0-30.4- 70	0-29.7- 90	5,2,5, 6, 6
84 " "		20-60	0-72.5-100	10-50.0-100	0-32.5-100	
56 days wet			10-52.0. 95	10-50.0- 80	0-53.7-100	
84 " "			10-58.8-100	20-72.1-100	0-57.5-100	-,-,4, 7, 6
Evernia prunastri						
56 days dry	-	0- 2.5- 15	0- 0 - 0	0- 3.5- 5	0- 0 - 0	
84 " "		5- 4.6- 17.5	0- 0 - 0	0-15.5- 50	0- 0 - 0	-,6-4, 5, 5
56 days wet	-		0-12.4- 30	0- 3.0- 5	0- 7.5- 15	-,-,4, 5, 5
84 " "			0-15.0- 40	0- 3.0- 7	0-26.5- 50	
Parmelia physodes						
56 days dry	0-0- 0	0-23.8- 95	0-49.4- 95	0-30.8- 90	0-50.3-100	7,4,8, 9,10
84 " "		0-25.9-100	15-81.3-100	10-65.6-100	0-53.0-100	
56 days wet			0-57.4-100	5-47.8-100	0-55.3-100	-,-,7.10,10
84 " "			5-68.4-100	10-55.0-100	5-68.7-100	
Parmelia saxatilis						
56 days dry	0-0- 0	30-76,7-100	15-43.8-100	0-24.6- 80	0-30.5- 60	5,3,4, 7, 5
84 " "		30-76.7-100	20-75.0-100	33-72.6-100	0-34.2- 60	
56 days wet			20-37.0- 70	15-48.3-100	20-73.6-100	-,-,5, 6, 7
84 " "			10-49.5- 95	20-60.0-100	35-75.0-100	
Parmelia sulcata						
56 days dry	0- 0	50	0-33.3-100	0-10	0-46.3-100	2,1,3, 2, 4
84 " "		80	20-40.0-100	50-10	5-47.5-100	
56 days wet			0-38.3-100	30-30	20-100	-,-,3, 2, 2
84 " "			10-70.0-100	40-50	70-100	

Experiment 2. In each column the lowest average and highest percentages discoloured surface are given.
The 'wet' group was sprinkled regularly, the 'dry' group was not sprinkled.
Boxed average values indicate significant divergence from the control values.

Appendix 113

Species and sampling area	experiment number	number of samples per column	controls	O_3	SO_2	$SO_2 + O_3$	C_2H_4
	3 4			44 ppb 60 ppb	100 ppb 100 ppb	100+60 ppb	100 ppb
<u>Buellia canescens</u>							
Leersum	4	9-10	45-80.5-95	5-[54.5]-100	40-83.0-100	80-92.0-100	
<u>Evernia prunastri</u>							
Scherpenzeel	3	6- 9	0-.2.2-10	0-[6.7]- 12.5	0-[34.4]- 70		15-[52.5]-100
Maarsbergen	4	10	0- 8.5-25	0-[40.2]- 75	5-[38.0]- 95	25-[45.0]- 98	
<u>Parmelia physodes</u>							
Glimmen	3	8-10	0- 3.3-15	0- 0 - 0	25-[39.7]- 90		0-20.6- 80
Speulderbos	3	3- 5	0- 0 - 0	0- 0 - 0	5-[9.2]- 15		0- 1.3- 5
Leersum	4	10	0-25.2-60	0-[6.3]- 15	0-36.7-100	0-31.7- 60	
<u>Parmelia saxatilis</u>							
Speulderbos	3	8- 9	0- 1.9- 8	0- 0.3- 3	0-[14.4]- 25		0-[8.5]- 40
Speulderbos	4	9-10	0-36.0- 8	5-36.1- 90	0-39.5-100	0-35.6- 75	
<u>Parmelia sulcata</u>							
Glimmen	3	8	0- 4.1-15	0- 2.1- 5	10-[35.9]- 90		5- 5.3- 12.5
Leerdam	4	9	5-17.8-80	5-32.8-100	5-[28.6]-100	5-[34.7]-100	
<u>Physcia tenella</u>							
Glimmen	3	4- 5	0- 0 - 0	0- 2.5- 5	0-[35.0]- 90		0-[44.5]- 90
Langbroek	4	10	0-30.0-80	8-[62.8]-100	0-38.2- 80	7.5-[50.2]-100	
<u>Xanthoria candelaria</u>							
Leersum	4	10	0-36.0-100	0-46.0-100	0-33.0- 80	0-61.0-100	

<u>Experiments 3 and 4.</u> Duration 170 and 180 days, respectively. For exact concentrations, see Appendix 110. In each column the lowest, average and highest percentages discoloured surface are given, Boxed values indicate significant divergence from the control values. For <u>X. candelaria</u> the percentages represent the amount of thallus lost, rather than discoloured surface, because of detachment of dead thallus in this species.

Regions:	
Biesbosch	M
Gaasterland	D
Noordoost polder	F
de Peel	O
Texel	B
Twenthe	G
Veluwe	H
Voorne	L
Wadden	A
Westland	K
Wieringermeer	C
IJsselmeer	E
Zeeuws Vlaanderen	N

towns, villages etc.:	
Aalsmeer	10
Amsterdam	9
Arnhem	26
Beverwijk	8
Bilthoven	20
Bleiswijk	16
Delft	15
Dordrecht	28
Eindhoven	30
Glimmen	3
's-Gravenhage	13
Groningen	1
Haren	2
den Helder	4
Hoek van Holland	14
Hoorn	6
Langbroek	21
Leerdam	29
Leersum	23
Leiden	12
Maarsbergen	22
Maastricht	31
Nijmegen	27
Rotterdam	18
Scherpenzeel	24
Schiedam	17
Schokland	7
Speulderbos	11
Utrecht	19
Wageningen	25
Wijster	5